W9-BMY-499

Norm Abram's New House

ALSO BY NORM ABRAM:

Measure Twice, Cut Once

The New Yankee Workshop Outdoor Projects

Mostly Shaker from The New Yankee Workshop

Classics from The New Yankee Workshop

The New Yankee Workshop

Norm Abram's New House

NORM ABRAM

Photographs by Richard Howard
Illustrations by John Murphy

Little, Brown and Company
BOSTON   NEW YORK   TORONTO   LONDON

Copyright © 1995 by Norm Abram

All rights reserved. No part of this book may be reproduced in any form or by any electronic or mechanical means, including information storage and retrieval systems, without permission in writing from the publisher, except by a reviewer who may quote brief passages in a review.

FIRST PAPERBACK EDITION

Library of Congress Cataloging-in-Publication Data

Abram, Norm.
    Norm Abram's new house / Norm Abram. — 1st ed.
      p. cm.
    Includes index.
    ISBN 0-316-00487-1 (HC)   0-316-00410-3 (PB)
    1. Abram, Norm. 2. Carpenters — United States — Biography.
  3. House construction. I. Title. II. Title: New house.
HD8039.C32U615   1995
690'.837'092 — dc20   95-2744

10 9 8 7 6 5 4 3 2 1

MV-NY

Published simultaneously in Canada by Little, Brown & Company (Canada) Limited
Printed in the United States of America

TO MY WIFE,

LAURA, *who patiently let me fulfill my dream, and*

TO MY DAUGHTER,

LINDSEY, *to whom I owe many attentive weekends to make*

up for all the time I spent on the new house

Contents

Acknowledgments

SHORTLY AFTER I TOLD him I intended to build a house, Russell Morash, my producer on *The New Yankee Workshop* and *This Old House,* encouraged me to write this book. I think Russ believes every enterprise should be converted into at least one other medium. He is a genius at doing it himself. William Phillips at Little, Brown offered me a contract without hesitation, my fifth with him and his house. Catherine Crawford edited the manuscript with care, occasionally imploring me to write faster; but you know me: think at least twice before writing down once. Pamela Marshall copyedited the chapters, as she had previously copyedited some of my woodworking books, thoughtfully. It was my responsibility to deliver a manuscript and illustrations to the publisher. I couldn't do it all myself. Bob Payne helped me formulate my first approach to the book, and Terri Kahn kept track of the first phases of construction. Don Cutler, otherwise my agent, helped me with the final stages of the book. Richard Howard came to the site many times to document one phase after another photographically. John Murphy, who does the final drawings for my woodworking books, made the drawings requested by Pamela Hartford, my art director, who also selected the photographs and assisted with captioning. The look of the book is attributable to their collaborative work and to the book's designer, Caroline Hagen. They understood intuitively that I wanted the book to look as traditional as the new house. Thank you, all of you.

Preface

EVERY HOUSE HAS ITS own story. The story begins as someone's dream. The dream becomes a plan, maybe of a completely new house or maybe of an existing house renovated to give it new character. Craftsmen and tradesmen of many skills conspire to follow the plan and make a new house or a new wing of an old house rise out of the ground, usually not without some surprises and changes along the way. Over the years, as a craftsman, I've helped fulfill the dreams and follow the plans of hundreds of other people's houses.

This book is based on the premise that when I finally built the house of *my* dreams, my friends might enjoy reading its story. What manner of house would my wife, Laura, and I choose to build? What decisions would I make about its structure and the materials used? Would I experience the typical frustrations of others — delays, unforeseen problems, a certain amount of disappointing workmanship — and how would I react to them when it was my house?

Telling the story of building a house has presented me with one overarching challenge. Fundamentally, the story should be told chronologically because that's the way we all like stories to be told — with a proper beginning, middle, and ending. Building a house, however, does not follow a simple chronological progression. The first stages are relatively cooperative. The house is designed, the site cleared, the foundation excavated; not too many things are happening at one time.

Planning	1990 J F M A M J J A S O N D	1991 J F M A M J J A S O N D	1992 J F M A M J
Shopping for a house			
Shopping for land purchase			
Siting and House Design			
Conservation Commission			

But then the plot thickens. On a single day, one subcontractor is applying shingles to the roof, another is nailing clapboards on the facade, yet another is building a chimney; and I am consulting with the architectural designer about still another aspect while Laura shops for bathroom fixtures. The story line loses out to the competing subplots if everything is told at once. Or the painter comes and does a couple of days' work, then disappears for a week or two, returns, disappears again . . . So some aspects overlap too much, and some aspects get spread out over too long a time for good storytelling.

My strategy for preventing this complexity from muddying the story line is to divide the story into chapters, each of which discusses one aspect or a few interrelated aspects of the process and follows them pretty much from beginning to end. In general the chapters move ahead chronologically from one to the next, but sometimes there is a doubling back in time from the end of a chapter to the beginning of the chapter following. I've tried to give enough time references to keep the narrative clear while recognizing that it is the story that counts most, not the actual dates.

It occurred to me after I finished the manuscript that it might be helpful to give an overview of the time line of building the house near the front of the book as a guide to the more detailed narrative that follows. I devised the accompanying chart for that purpose. You can see how long it took us to get from first thoughts of another house to actually breaking ground for a new house. I don't regret the slow takeoff; the moral is that planning is critical and shouldn't be rushed. You can see the process get thicker with overlapping pieces of the construction. You can see what little blips on the screen some of the pieces occupy, how extended the time line bars for other pieces are. I

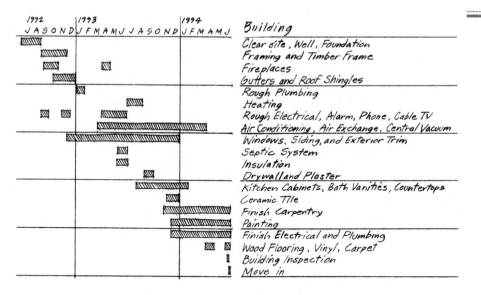

hope referring to the chart will be helpful if you get confused about the relation of any phase to the whole process.

In the course of reading this story you will meet experts in many aspects of residential construction, They, too, have stories; they have faces. Most of them are self-employed individuals, some have partners, a few are from family-owned businesses. I've tried to give enough of a glimpse of them so that both their talents and their temperaments are recorded. They are a microcosm of the best of a whole industry. You can't build a good house without them.

This is a story with a happy ending. The beginning, though, is — as we say — "another story."

A Crisis

FEBRUARY, 1991. I WAS standing on a stepladder in the attic of a spacious three-and-a-half-year-old Cape Cod–style house located about twenty-five miles west of Boston, Massachusetts. Flashlight in one hand, my clipboard in the other, I reached out in the dark to touch the plywood underside of the roof exposed between two rafters. It was damp. Another problem, another note to add to the lengthy list already jotted down on my clipboard. Was this house worth its $600,000 asking price? I had to decide in the next few minutes; a real estate broker was waiting downstairs for me to make an offer.

Laura was waiting with the broker. She had fallen in love with this house the moment she first set foot in it.

When it comes to houses and furniture, and just about everything else, Laura and I have very similar tastes. We like traditional over contemporary, comfortable over showy. There was a lot to like about this house. It stood on a rise at the end of a private cul-de-sac, surrounded by sweeping lawns begging to become a country garden. The house was pretty enough to be an illustration for a Christmas card. Inside the house there were wide plank floors and Rumford-style fireplaces. Best of all, in Laura's eyes, was a cozy, fireplaced sitting room with built-in shelves for books and her collections of porcelain dolls and antique plates.

For an hour I had been going over the house by myself, looking past its obvious charm to examine its construction. I wasn't leaving any marks of my investigation — the owners, fortunately absent, would never know how closely I was inspecting their home — but the real issue was whether the house was making the right mark on me. Was it built well enough to meet the standards I would impose if I were building a new house for our family? Was the exterior of the house protecting its interior from the constant stress of New England's climate? The builder had a good reputation, but from my first approach to the house I began to see problems.

Walking around the exterior, I had noted on my clipboard that nails were working loose on the siding and around the window frames of the south wall. Had the carpenters missed the studs with those nails? Did they use the wrong nails? By its very nature, wood expands and contracts — particularly across the grain — as its moisture content changes. Moist air, which causes the wood to swell, doesn't come just from exterior weather conditions; it can come as well from moist interior air reaching the interior surface of the siding. Extreme heat beating against the south wall in summer would cause maximum shrinkage of the siding after it had swelled in preceding seasons. Under those alternating conditions of first swelling and then shrinking, improperly installed siding would begin to pop its nails.

My hunch was that the builder hadn't primed the back side of the siding with paint or stain, a step that allows the siding to absorb moisture more evenly. Or the builder may have poorly installed a moisture barrier on the inside of the stud walls. If I owned the house, I would renail the siding, being sure to nail into the studs, and watch it for a while; if the nail-popping recurred, I could tear off the siding and replace it with back-primed siding; but that would not solve a problem caused by an improperly installed moisture barrier.

In order to give the house as authentic a Cape-style look as possible, the builder had brought the siding down very close to ground level, in some places within three or four inches of the ground. The state building code in Massachusetts calls for a minimum of six to eight inches of clearance between soil and any wood used in exterior construction. In an effort to prevent soil from being too close to the sills and siding, the builder had surrounded

the perimeter of the house with a twelve- to eighteen-inch-wide border of crushed stone. The stone border was meant to keep damp organic material away from the house and absorb surface rainwater quickly, keeping the sills and siding dry to prevent rotting and insect infiltration. However, there were many grasses and weeds growing up through the stone border, which made me apprehensive about the effectiveness of the border.

The attached garage showed an even worse problem. The vertical pine board siding on the garage ran right into the earth, an open invitation for the boards to soak up moisture and offer termites or carpenter ants a good home. How could the local building inspector have permitted it?

By the time I got to the attic my clipboard already contained a sobering list of problems and reservations. There I discovered that the wood-shingle roof had been laid directly onto the plywood sheathing covering the rafters. No provision had been made for air to circulate between the topside of the plywood and the underside of the shingles. This roof could need replacing within five years.

The eaves in a Cape Cod–style house come down to the first floor, making the upstairs a half-story. Laura's love was an oversize, contemporary, Cape-style house with dormers to enhance the space upstairs.

Another thing I noticed was inadequate venting of the attic to remove trapped, moisture-laden air to the outdoors. This air was migrating into the attic from the living spaces below. Because the air was trapped, the plywood sheathing was damp where I was touching it between the rafters. On really cold days, I suspected, frost might form where this warm air interacted with the damp plywood. In summer this attic would get too hot, prematurely aging the shingles and causing the framing materials to shrink more than they should, maybe loosening the grip of framing nails.

I could imagine myself out on the roof of this house in mid-February chipping away at ice, because it was evident that a lot of warm, moist air was escaping from house to attic. As snow accumulated on the roof in winter, the constant warm air rising to the attic would cause freeze-thaw cycles leading to ice damming on the roof. The roofing shingles would be no barricade to water backing up underneath them. So a sudden warm spell and/or winter rain could flood the attic floor and the ceilings and interior walls below.

With each additional notation of a problem I became more convinced that this wasn't the house for us. We were looking for a new home where we might live for decades. I wasn't sure I could live contentedly in this house for twenty-four hours.

Laura and the broker were waiting downstairs for me to come down and discuss our first offer on this house Laura had fallen in love with. But the house had failed my examination.

What was I going to do?

The Dream

IN THE SPRING of 1990 Laura and I began to talk about moving again. We had been married for eleven years and had lived in three different houses. After renting a one-story ranch-style house and a modest Cape Cod–style house, we bought our first house, a garrison colonial, in 1984. I remember what the garrison looked like when I first saw it. The yard was overgrown with weeds. The garage was filled with garbage. Four young men shared the house just before we bought it. Their social life apparently revolved around the full-size bar in the living room. The house smelled so much like a tavern that Laura brought along a can of room freshener to spray each room as she took me through.

Even so, I quickly saw what Laura had seen underneath the shabby surface when she scouted the house. We liked its style. It offered us the additional space we needed for our family in a quiet location on a safe, dead-end street. Its shabbiness brought the asking price well below what the house would have commanded if in better condition. Our Yankee sense of thriftiness appreciated getting the house at a bargain price.

During our first six years in the garrison, we improved the kitchen and the master bathroom, replaced the roof and gutters, installed central air-conditioning and a new furnace, added insulation, upgraded the electrical system, and landscaped the yards — plus the usual painting and redecorating. But there were still several major improvements we had in mind that I hadn't yet gotten to, for example, expanding the master bedroom on the second floor and adding a family room and deck underneath it.

Being a carpenter and contractor, I did most of the renovation myself, not always on the fastest possible schedule. Much of the time there was at least one area of the house torn up and only slowly getting put back together the new way we wanted it to be. Skepticism about my pace on projects at home was the principal reason we first began talking about *moving*, not about *building*. Laura once said to a friend that if I were to build a new house at the same pace that I had renovated our present house, "I might kill him."

Why did we want to move at all? One reason was to find an excellent school for our daughter, Lindsey, who would soon be in junior high school. Laura's other three children — my stepchildren — were grown up and on their own. Lindsey is a bright, self-motivated student, and we wanted the best possible school for her. The schools where we lived were all right, but closer to Boston we could choose among regional schools ranked with the best in the state.

Moving a few miles in a northeasterly direction would give me an easier commute to the workshop-studio where I design and make furniture for *The New Yankee Workshop* on public television. I'd be nearer, too, to the Boston-area home renovation projects that we televise on *This Old House*. Sometimes *New Yankee Workshop* and *This Old House* are in production at the same time, and we switch from one series to the other day by day — or even from morning to afternoon. Many nights I work late in the workshop refining a new piece of furniture. Shortening my commute would make my workday just that much easier.

When Laura and I began talking about moving to another house, we also took into consideration where we are in our lives. It was the right time, we believed, to move into the major house of our life, a house that would shelter us together through grandparenthood and on into retirement years.

We could have just the right amount of space — not too little, not too much — for children and grandchildren to visit. Since I do most of my office work at home, I wanted a better office than the one I was currently using, which had been fashioned out of a spare bedroom. I need a place away from the main rooms of the house where I can sketch furniture designs, make

measured drawings, work on book projects, and keep records on my computer — but also not be bothered by the home office telephone when I am
enjoying family time.

Laura needed a better place to house her tropical birds. She first acquired
a few as a hobby, but the hobby was beginning to look as though it might
become our extended family. The parrots needed a room of their own, too.

What would the house of my dreams look like? Since I first began working
with my carpenter father in the summer of 1965, I have built houses to fulfill
many people's dreams. I knew by now what I would most like to build for my
family, following the example of my father, who built the house of his dreams,
in which he and my mother still live. I've spent my life in New England and
I've always loved wood-frame, two-story colonial houses. My dream home
would have a classic center-entry facade with double-hung, divided-light
windows and a wood-shingle roof and interior and exterior features that require minimal maintenance, take advantage of recent technological innovations, and complement a colonial design.

The demands of my schedule and the condition of the real estate market
in 1990, however, made me hesitate about building a house myself. Housing
in New England had not yet begun to recover from the dramatic decline in
value caused by the recession of the late 1980s. I thought we might find some
bargains in recently constructed houses. If there was an existing house for
sale whose style and location appealed to us and whose quality of construction met my standards, I believed I'd be willing to set aside my dream of
building it myself.

Laura began to scout houses for sale. She walked through fifty or sixty
houses during the next several months. On Sundays we drove by houses she
had seen recently, and while I looked at the exteriors Laura described the
interiors to me. Or we scouted exteriors together in drive-bys, and when we
found a house that suited our taste Laura made an appointment to see the
interior.

The fact that I am a television celebrity was a hindrance. Realtors assumed
we would want something grand and flashy: marble foyers, even servants'
quarters. "A marble foyer isn't me," Laura said. "It's too cold. I want warmth.
I want a house we can be ourselves in." Finally, after she had visited more

than thirty houses, none of them remotely interesting to her, she turned to the broker in frustration and said, "Look, we're really country people."

In February 1991, when our search was almost a year old, a realtor named Ellen took Laura to see a house she fell in love with. The oversize Cape Cod–style house had been on the market when Laura first began scouting, but Ellen at that point was sure we were looking for glitter. Laura showed me a photograph of the house one Thursday night. On Friday morning we took a quick walk through the house before I left for a weekend personal appearance; Ellen made an appointment so that I could go over the house carefully the following Monday.

In a more typical case, a couple who found a house they liked would engage an engineer, a home inspector, or a contractor to go over the house carefully and evaluate its condition. Because of my experience as a contractor, I could do the evaluation myself. Even on the quick Friday walk-through, I began to catalog some problems in my mind. But I didn't want to commit myself to an opinion, particularly when Laura was so enthusiastic. "Well, I don't know," I said. "I'll take a closer look on Monday."

When I called home from my road trip, Laura pressed me for my appraisal

I like the simplicity and symmetry of the eighteenth-century houses still abundant around Boston. They say "home" to me. Framed and clad in wood, my favorite material, they are easy to copy and adapt for contemporary houses.

of the Cape-style house. That weekend she drove past the house several times, and when I got home she had mentally furnished every room and was ready to move in immediately. I've always had some misgivings about the Cape Cod design for a house. Its steeply pitched roof doesn't foster the best interior space arrangements, especially on the second floor. Its rafters and knee walls bracing the rafters are difficult to insulate and ventilate. But those factors weren't what made Laura fall in love; she fell for the location and obvious charm of the house, and because she felt "at home" when she walked into it.

All that was on my mind as I came down from the attic. Laura and Ellen were waiting in the garage, and as I approached them I just shook my head. "I don't think so," I said. "If we offer what I think this house is really worth, I'm just going to insult the owner." (I was ready to knock $250,000 off the asking price.) Laura was speechless; she hadn't noticed the flaws, many of which were hidden by the finish work.

Ellen, beginning to see a sale and its commission evaporate, urged me to make an offer, any offer. "Why don't you go to lunch, then swing by my office?" she suggested when I was silent. At a nearby coffee shop Laura and I had, I believe, our most miserable lunch together ever. I tried to explain to her what was wrong with the construction of the house. "Just fix it," she begged when I described the roof. She was close to tears.

But I knew I wouldn't. I also knew I was breaking Laura's heart.

We drove back to the real estate office in silence. Ellen had talked to her boss, and again asked us to make some kind of offer. "I can't even do that now," I replied. "I've had a couple of hours to think, and I'd never be happy there. I'd be miserable. I don't think that's the house for us." We left the office as silently as we left the coffee shop, climbed into the Bronco, and then the tears flowed.

For several days Laura and I couldn't talk about what had happened. I couldn't console her because I was the bad guy who had made her sad. She kept a photograph of the Cape-style house and would cry sometimes when she glanced at it. "Losing the house I wanted," she said to me months later, "was such a letdown that for a while I had a hard time thinking about a new home at all. I was afraid to get excited again." During the fateful meeting in

Ellen's office I suggested to Laura that we look for land to build on rather than for an existing house, but it was not a good moment to make a new proposal.

A few weeks after our tense disagreement over the Cape-style house, I was on the road again for a personal appearance. Sitting in my hotel room I came to a decision. Hectic schedule or not, I would take time, find time — *make* time — to build us a new house. "Look," I said to Laura when I got home, "I can build you three times the house — with workmanship I'm comfortable with — for the same amount of money they were asking for the Cape. Give me a list of everything you want in a house. I'll make sure you get it. And I'll be happy building it."

Looking for undeveloped parcels of land proved to be different from shopping for houses. We didn't need any appointments with homeowners. All we needed were clear directions to prospective properties, then adequate descriptions or plans so that we could discern their boundaries. There were four towns whose school systems interested us. We visited Ellen at her real estate office again and asked to see listings for them. In the four towns, there weren't more than a dozen parcels that met our specifications: at least two acres in a location that offered plenty of privacy. Both Laura and I grew up in New England towns surrounded with farmland outlined in old, hand-built

The "ell" (short for elbow) extending behind this early nineteenth-century house, connecting house to barn, is a form we borrowed to link our house to the office/garage wing. As the owner of this house felt free to modify the more private ell with modern glass doors, so we felt free to build an ell with a south-facing glass wall.

stone walls. We wanted land with a rural feel to it. I wanted a site that promised to hold its investment value in the years ahead, but I also wanted to avoid pretentious areas where nearby owners had spent more on stonework and landscaping than we planned to spend on the house itself. (We saw a few of those.)

Both Laura and Lindsey were with me one wintry afternoon when I drove our Bronco around the rotary where three roads converged at the village center of one of the four targeted towns. Snow covered the ground and sat in little caps on the granite fence posts lining the edge of the rotary.

The village center itself looked rural. Town offices and the town library shared a turreted brick hall. The police operated out of what looked like a Cape-style house; a branch office of a regional bank occupied another Cape. Next door to a small gas station at the rotary, in a clapboard building, a mom-and-pop business offered newspapers, basic groceries, and some excellent deli items for sale. The dozen or so houses located close to the rotary all looked to be of nineteenth-century vintage or older. There were no mini-malls, no chain stores, no fast food restaurants. We could have been in rural Vermont or New Hampshire. No doubt about it, this was a quiet town I could happily settle into.

Less than two miles beyond the rotary, following written directions, I turned off the main road into a subdivision with plots in the two- to four-acre range. The subdivision street curved and twisted in such a way that the several houses already built there didn't front directly on one another. Most of the houses were at least vaguely colonial in design.

From the subdivision street I turned onto a long private drive, which was the access road to four plots, three of which had already been built on. The farther we went along this drive, the more wooded and private it became. I parked next to a stone wall at least a century old. Laura, aware of the bite of the day's fierce north wind, elected to stay in the warm Bronco, but Lindsey was willing to go exploring. We clambered out of the truck and over the wall. Quickly we were in a thick stand of pine and oak trees. A new house stood on another parcel about a hundred yards from where I had parked, but once Lindsey and I had gone a few yards into the property for sale we couldn't see the neighboring house anymore.

Running across the property on several tangents were other old stone walls, one with a break in it, where I suspect an old cart path once passed through. Near the opening in the wall we came upon the open, five-foot-deep foundation of a small building no longer standing. The rectangular hole was lined with huge, roughly hewn pieces of granite. Because of other pieces of quarried granite lying in the vicinity, I thought the foundation might once have supported a storage hut for a quarrying operation.

I had the realtor's small photocopy of the plot plan with me, but it wasn't very helpful in determining the boundaries of the property we were tramping through, and the weather was daunting, so Lindsey and I didn't explore very long. Back in the warm Bronco, I looked at Laura and said, "Hmm. I suppose it has potential."

Ellen, who had suggested this property to us, returned with Laura and me the following Saturday in milder weather. We met the owner of the property and a second realtor, Maureen, who, as the original or listing agent for the sale of this piece of land, had to be involved in its sale. Our meeting with the owner was very brief, and we never saw him again. He left with me a large-scale (1″ = 40′) plot plan showing boundaries, directions, the location of the septic system, for which the local Board of Health had given approval, and the location where he had planned to build his house before he changed his mind and put the plot up for resale. Naturally I wondered what had caused him to change his mind about building there, but I didn't ask.

Laura and I began a more thorough exploration of the plot, looking for the boundary markers indicated on the plot plan. The piece of property — a little more than four acres in size, one of the largest in the subdivision — was long and relatively narrow. Its long axis ran from east to west, beginning where the access drive intersected the subdivision street. The access drive ran through the east end and then along the north side of the parcel. On both our first visit and this second one, I had driven to the back third of the parcel to park. The east end of the property, out by the street, was marked as wetland, where building was prohibited, but I wouldn't have wanted to build near the street anyhow.

The plot was shaped somewhat like a bow tie. By the center knot of the tie the plot was narrowest, and a little beyond the knot, to the west, the plot reached its highest elevation on a knoll dominated by a large rock outcropping. The owner had planned to build his house on top of the knoll. The long, straight south boundary faced a forest that was not part of the subdivision. Who owned it? Would it become a developed subdivision in the near future, with houses close to and visible from any house we might build there?

Walking on from the knoll toward the far, or west, end of the property, Laura and I found that the land flared out to greater width and sloped off rather sharply. Near the western boundary there was an embankment, at the foot of which I could see water seeping up through the dark earth to become a small stream. Was this wetland too? It was not so marked on the plot plan.

The plot was certainly an unusual shape — low and wet at both ends, higher and dry in the middle — but nothing we had yet found in the four towns matched it for privacy or for the natural beauty of its woods and walls.

Laura and I immediately began to talk about where a house could best be positioned. We both doubted the wisdom of putting it where the seller had had in mind — on the very top of the knoll — because a house located there would look directly down on the house of the nearest neighbor. I thought perhaps a house could be sited toward the back of the knoll and drop off the edge with a garage at a lower level coming in from the back.

"Why not put a house on the slope just behind the knoll?" Laura suggested. I was pleased that she was beginning to imagine *our* house on that land, but I initially resisted her suggestion. The slope did have a good exposure to the south overlooking the forest next door, and the knoll would protect a house from the harshest winter winds from the north. But the knoll would loom close to the north side of the house. "Would we be happy looking out one side of the house at the back of a hill?" I replied.

I was also concerned about the rock outcropping on top of the knoll and the evidence of granite quarrying on the property. How much rock lay below the surface where one might want to dig the foundation of a house? How much expensive blasting might be required to wedge out an adequate basement?

The potential cost of excavating through rock made me hesitate and look at a few other pieces of undeveloped land, but at the same time we continued to make visits to the bow tie–shaped plot. I also made an excursion to the town hall to check the tax records on each property adjacent to it. I was re-assured to find that the forest on the long south boundary was part of a fifty-acre tract held in trust. The tract didn't seem to be a candidate for development in the near future, and it didn't look like a tract that would be easy to develop. Maybe the owners would designate it for conservation. I decided I could take my chances on the fate of the forest, even though its edge came pretty close to any house I might build.

Without finally deciding where the house would be sited, but knowing there would probably be high excavation costs no matter where we sited it, Laura and I decided to attempt to buy the parcel. We liked the setting enough to be willing to deal with any construction problems.

The two realtors, Ellen and Maureen, were working mainly for the inter-ests of the seller, but Ellen was willing to give me her guess as to how much less than the asking price the owner might accept to close a deal. I thought her figure was too low, so I raised it to what I hoped would be acceptable, made an offer, and the owner rejected it. I then suggested that the owner make a counterproposal, but he declined to do that. So I split the difference between our positions and made a second offer.

The owner rejected the second offer, too. We were separated by a couple of thousand dollars. I decided to be firm about my second offer being my best offer. Laura was nervous that we might lose the property over a compara-tively small difference, but the realtors took my firmness seriously. They hud-dled and decided to absorb the difference by taking a smaller commission in the interest of closing the deal.

I proposed as a condition of the sale that the seller acquire for us from the Board of Health an extension of his permit for a septic system; it was almost three years old and about to expire. The seller refused to do anything about the permit. Either he didn't want to take the time to go down to the town hall or he didn't want to pay the $50 fee for the extension. My attorney arranged to have it extended.

While the seller and I negotiated through our intermediaries, my lawyer, Frank Morris, was busy searching through the Registry of Deeds to make sure the title to the property was clear — that is, that there were no disputed claims of ownership or liens (financial claims) against the property. The county records went back over a hundred years. In spite of the absence of any troubling evidence, Frank recommended that we buy title insurance, protecting Laura and me for the value of what we were acquiring (but not for the value of what we might build on it) from any surprise claims. In Massachusetts title insurance is optional, but I had only to look at something like the abandoned granite foundation to know I didn't want to take any chances that a long-lost heir might come forward.

It was time to bring in our first consultant, a landscape architect. Many people, I believe, bring a landscape architect into the process much too late, sometimes not until after the house has been built. I advocate bringing one in right at the beginning to help position the house wisely in its natural setting, to recommend the path of driveways, and to suggest which trees should be cleared, which spared. I called Tom Wirth.

I've known Tom since 1980, when we worked together on the second project televised on *This Old House:* renovation of the massive (and massively rotted) Bigelow house in Newton, Massachusetts, designed by H. H. Richardson. By the time we had set all of the grades, steps, and walkways around the mansion's five new condominiums, I had come to admire Tom's sensitivity to the natural orientation of a house. He likes, for example, the way colonial New Englanders backed their houses against natural shelter when they could, facing their buildings to the south to catch the sun.

Tom's choice of profession was settled early in his life. His grandmother gave him his first garden patch at age six in rural Bucks County, Pennsylvania. He grew up through 4-H Club activities, which took him as a high school student to Penn State for a vegetable-judging contest; the point was to determine the extent to which student judges' opinions would match a professional judge's opinions. Tom doesn't remember how well he did. The contest was held, coincidentally, at the university's landscape architecture department, and he was distracted from the judging contest by the landscape architecture drawings and photographs on display. He had never seen anything like them. A professor from the department, seeing his interest, engaged him in conversation. A career was born.

Laura and I held our first meeting with Tom at the workshop-studio where we videotape *The New Yankee Workshop.* We described to him the basic

"footprint" or perimeter of the house we had already agreed upon. It was to be a house in three distinct but connected sections. The front section, about forty-five to fifty feet by thirty feet, was to look like a two-story house of classic colonial design. Soon we began to refer to this section as the "main house." From the back of the main house would extend a twenty-two-by-thirty-six-foot one-story ell containing a kitchen and family room. (An "ell" is a secondary wing or extension of a building at right angles to its principal dimension — short for elbow.) Another two-story section for my office, the garage, and other rooms was to sit behind the ell, parallel to the main house, so that the whole structure exhibited the shape of a somewhat irregular capital **H**.

To determine the dimensions of the main house, I took the approximate size of our current house, enlarged the rooms just a little, and added space for the center hall our house lacked. I also looked for comparison at the plans for other colonial houses. For the office wing at the back, I began with the standard size of a two-car garage — twenty-four feet square — which the

We don't know what this old granite foundation supported. Tom Wirth hoped to plant a garden in the twelve-by-eighteen-foot depression. Laura was uncertain about it.

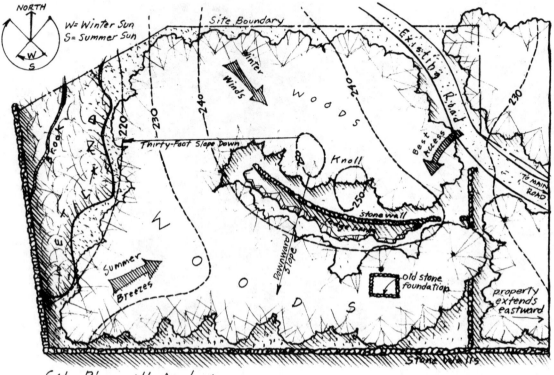

Site Plan with Analysis

wing would incorporate on its lower level. Then I added space for other functions, such as a utility room, a laundry room, a bathroom, and an exercise room. The size of the ell between the other two sections was determined by thinking through the amount of kitchen and family dining space we wanted and then doubling the length of the kitchen to provide space for a family room next to it.

Anyone who has driven through New England and seen the countless old houses in which the main house is joined by one or (often) more intermediate sections to the barn knows the inspiration for our dreaming of one house with three connected sections. In due course I intended to have another building near the house for my own woodworking shop, but I was uncertain whether we could build both house and shop at the same time.

Tom Wirth listened carefully to Laura and me as we talked about our dreams for the house. Later he showed me his notes. He had jotted down key words and phrases such as "privacy," "country informal," "herbs and perennials," and "simple." Since we were meeting in June 1991 and planned to build something "simple," Tom also jotted down by way of schedule, "Break ground, fall. Framed by Christmas." I report this latter note just to show how unreasonably optimistic we all were at this stage of the venture. We got the house framed just in time for Christmas, all right, but not the Christmas Tom had in mind!

Then the three of us drove from the studio to the site. Tom had designed the landscape of a house just two miles from our property and knew the area and its possibilities and challenges. He told us immediately how much he liked the site and its stand of hardwood and evergreen trees. The big chunks of granite with quarrying scars lying about delighted him. He quickly announced his vote on the issue of the best location for the house. Looking at the slope behind the knoll, just where Laura had suggested on her first visit, he said, "This is the spot for the house."

In New England, Tom recommends facing a house, wherever feasible, toward the southeast/south/southwest to get the most sun. On our slope, however, he recommended facing the house to the east toward the access drive. A house so situated gets morning sun on its front. One side of the long sweep of main house–ell–office wing would face to the south for day-long sun; the other long side would be nestled against the knoll for protection against the blasts of colder weather.

Tom's recommendation meant that the old open foundation with its rough granite walls would be in the front yard. The abandoned cellar hole reminded him of sunken gardens he had seen in Bucks County as a child. Laura wasn't happy with the prospect of keeping the exposed foundation. It looked creepy to her. "Just fill it in," she suggested. "Fill it in?" Tom repeated. "Oh, I hope you don't. It'll make a natural English garden."

I gave Tom a few pencil sketches I had already made showing possible relationships of the main house, the ell, and the office wing to one another. One of my sketches showed the ell coming straight back from the main house, but another had the ell coming off at a sixty-degree angle.

In a few weeks, Tom sent back three sketches of his own. His first seemed clearly the right one. The ell was centered on the back of the main house and came straight back from it. The office wing was at a right angle to the ell but not centered against the ell; it projected farther south than the main house did. Very simple, very New England. The separate building for my shop would be built to the west of the house, toward the far end of the property.

One aspect of Tom's sketch bothered me, however. He saw that the slope into which the house would be set fell away from east to west as well as from north to south. The east-to-west decline wasn't as prominent as the north-to-south decline, but it was there, and it would naturally make the ell and office wing sit lower than the main house. The way to deal with the change in elevation, Tom proposed, was to have a two-foot (or four-step) set of steps from the center hall of the main house down to the floor of the ell. I didn't like the idea of the step down. I wasn't sure at the time how to manage it, but my vision of the house was that it would sweep through on the same level from the front door all the way to the far end of the ell.

The driveway we all agreed on comes in from where I parked the Bronco beginning with our first visit. It passes through the opening in the stone wall made for the old cart path and curves around in an S to bring visitors to the front of the house, where they can park near the sunken foundation. For the family, the driveway continues along the south boundary of the plot past the ell and curves into a wider parking area outside the garage.

What I liked most about the siting of the house and driveway was the element of surprise. Only a small part of the top of the main house would be visible from the access drive. The driveway would bring a visitor around the knoll and in toward a colonial-vintage house. The ell and back wing probably wouldn't be visible yet. The house wouldn't reveal all of itself at first glance.

So far, so good. Where we were thinking of siting the house, however, put the house at a lower elevation than the septic field that had been approved adjacent to where the access road and our proposed driveway met. The developer had done a good job of finding the best location on our property for a septic field, but the field was going to be quite a distance from where we were thinking of setting the house.

Board of Health regulations require that any septic field be a minimum distance removed from any well or wetlands on the property to eliminate the

There are so many rocks and boulders near the surface of land in much of New England that clearing the land for pasture or planting necessitated building stone walls. We don't know how the land by this wall of ours was used before second-growth woodland reclaimed it.

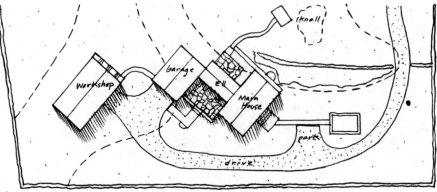

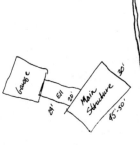

Scheme One

After I sent Tom
Wirth two rough
sketches of possible
footprints of our house
(above and below), he
drew three schemes.
We were drawn to his
version of my first
scheme. In it, the ell
comes directly back
from the center hall of
the main house.

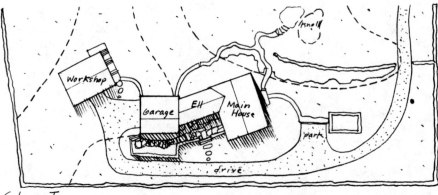

Scheme Two

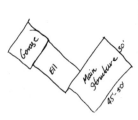

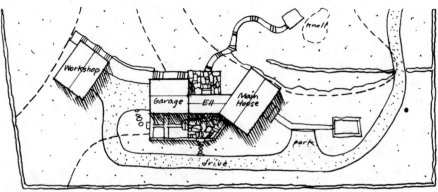

Scheme Three

contamination of either by waste seeping out of the septic field before it breaks down into harmless compounds. In our case, the septic field would have to be midway in the length of the property because the east end was definitely wetland and the well would be toward the west end, where, I suspected, there was also some undesignated wetland.

Usually, septic systems are fairly simple. The house sits at a higher elevation than the septic field. Waste flows by gravitational force from the house to an underground tank outside, where solids sink to the bottom (from which they can be removed as necessary by service trucks) and liquids drain into perforated pipes buried in trenches in the septic field, where the liquids can seep (or "leach") into the soil and be absorbed and broken down. If the house sits lower than the septic field, and gravitational flow therefore won't work, the owner can install a mechanical system to pump waste up to the field. But then one is adding equipment (and expense), which can malfunction and may eventually have to be replaced. A power failure would immobilize the system. The beauty of gravity is that it is free and flawless. Still, it looked as though we were going to have to invest in a septic pump.

Two other Board of Health regulations required the septic field soil to be porous enough to absorb liquid waste, and large and dry enough to contain the amount of waste likely to be generated by the residents of the house on the property. Porosity is measured by a percolation test, which the developer had already performed to get his permit. A civil engineer dug a hole at the specified location, poured water into the hole, and measured how quickly the water was absorbed into the surrounding soil. His test indicated an inch of water would be absorbed in five minutes, a more than adequate rate.

Probably with a backhoe, the developer performed the second test. He dug another hole at the septic field location to see if he could get to an acceptable depth before hitting water. The septic field has to sit a minimum distance above the water table at any time of year. He got down to seven and a half feet, again more than adequate, without hitting water or any other hindrance such as bedrock that would limit the area's septic capacity.

Since the seller had planned to position his house at the very highest point on the property, a simple gravitational septic system was not going to pose any problems for him. By relocating the house at a lower elevation, I was raising the probability of having to redesign the septic system but with the leaching field in the original location. If anything about the redesign required deepening the field, I would have to repeat the water table test. The percolation test can be done at any time of year, but the Board of Health requires the water table test to be done in spring after the frost is out of the ground but while the water table is at its highest. The testing period for 1991 was expiring. It looked as though certain questions about the septic system might not get resolved before the spring of 1992.

In the meantime, there was much to do. The parcel wasn't quite legally ours yet, and the house had been sited but not fully designed. The process of going from first offer to formal purchase of the land took about two months. The closing on May 28, 1991, was an anticlimax. Laura and I weren't even present. We had signed everything we needed to sign for our lawyer, and with some trepidation had written a check for the land larger than the purchase price of the house we were living in. Frank attended the closing and brought us back a folder of documents proving our ownership. Then Laura and I celebrated by going out to dinner. I thought back on the search we had conducted for several months with its occasional highs and its definite lows. I thought ahead to the designing to be done and several kinds of town permits still to obtain. And I remembered what my father had said one day when I was groaning about all of the paperwork. He smiled. "When your mother and I built our house," he said, "I just drove four stakes into the ground and called in the bulldozer."

"Country It Up"

FOR SEVERAL MONTHS in the summer and fall of 1991, after Laura and I had bought our four acres of land and were discussing details of our new house every day, I fantasized that I could build the house without engaging an architectural designer. If I could be my own general contractor, why could I not also design my own house? My mental picture of the house seemed so clear and detailed that I thought it would be easy to transfer the vision into all of the necessary working drawings. I know the basic techniques involved. I work with architectural plans all the time. And I keep myself in drafting practice by doing the preliminary measured drawings for the woodworking projects in my *New Yankee Workshop* books.

But time kept drifting away without my getting anything down on paper. Gradually I saw that I didn't have time to do the preliminary drafting and the inevitable revisions. I also realized that an architectural designer would undoubtedly bring ideas of his own to the house that Laura and I would endorse but wouldn't have thought of ourselves. The designer we consulted, Jock Gifford, pointed out a third contribution he could make. He could make sure that Laura's ideas and wishes didn't get overlooked in the course of planning the house I had been dreaming about for a long time.

There was nothing startling about our choice of Jock to help us design the house; it would have been almost shocking if we had gone to anyone else. He

was the designer when I did my first large job as a general contractor in the winter of 1976. A client of his was building a country store on the island of Nantucket, and I successfully bid on the job. Jock divides his time between Boston and Nantucket, where he owns a restaurant and other properties, and the large Gifford clan is prominent in island life. He has mastered adaptation of Early American architectural styles to houses suited to current standards of comfort. In the course of several years back and forth to Nantucket, I renovated parts of Jock's father's house and then I built Jock's new house.

While I was constructing Jock's house, he introduced me to his friend Russell Morash, who was preparing to build a small garage-shop near Boston, which Jock had designed for him. I put in a general contracting bid and got

Front entrances of colonial houses evolved from simply framed plank doors to doorways framed at top or sides with square or rectangular glass "lights." Ornamentation borrowed from Greek and Roman architecture in the Classic Revival often accompanied the lights. Our front entrance would have transom and side lights, a simple cornice header, and columns or "pilasters."

the job. As the garage-shop was taking shape, Russ approached me about doing carpentry on a new series he was creating for public television — inventing, producing, casting, and directing all at the same time — to be called *This Old House.* Jock Gifford would be architectural consultant on some of the houses renovated in the series. At first I wasn't sure it was a good idea. I'm cautious by nature, and I don't usually change the course of my life without thinking it over pretty carefully. Fifteen years later, I'm glad I made up my mind to accept Russ's invitation.

During our first meeting with Jock in November 1991, he outlined the exercises he asks all of his clients to perform. I knew he saw his task as helping to realize *my* and *Laura's* vision, not to design a house for us that fulfilled *his*

Jock Gifford, Jack Beruk, a window manufacturer's rep, and I study the specs in my New Yankee Workshop "office," trying to find the best match of colonial design and energy-efficient technology.

vision of a nice house. He wanted us to draw up a wish list of rooms and features we hoped to include in a new house. He asked us to indicate the general location of these rooms in a rough sketch called a "balloon drawing," which looks a little like elementary school art. Jock had already seen and studied the footprint and siting of the house that we worked out with Tom Wirth. He also had a line drawing I gave him of the facade of a house — as I recall, a house in Deerfield, Massachusetts — that represented the kind of colonial style we wanted to emulate. It was Jock's principal task to make our wish list fit as well as possible into the style and footprint we had decided upon before bringing him into the picture. Drawings and sketches began to flow back and forth between us.

It seemed natural to begin with the design of the main house. Laura and I envisioned the first floor having a spacious center hall running from the front door back to the opening into the ell. Perhaps a powder room and a coat closet could be positioned to open off the hallway. The stairway would rise conventionally along the right wall of the hallway, but — unconventionally

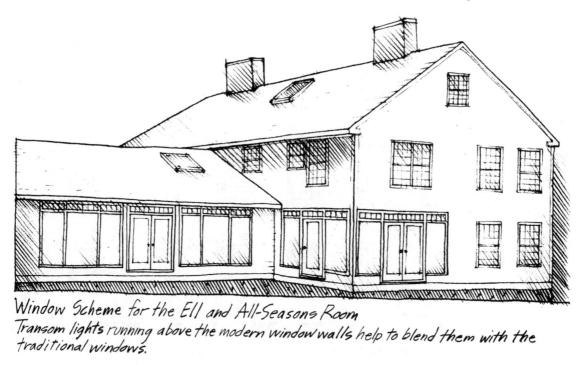

Window Scheme for the Ell and All-Seasons Room
Transom lights running above the modern window walls help to blend them with the traditional windows.

for an authentic early colonial house — we thought of having the hall open up to two-story height. Flanking the hallway we hoped to find room on the first floor for a formal living room, a formal dining room, and a library for Laura comparable to the sitting room she loved so much in the house we didn't buy.

When we were first discussing the footprint of the house with Tom Wirth, Laura mentioned her wish to have a screened porch attached to the new house. I couldn't imagine how a side porch could be incorporated and not look like an afterthought. Most screened porches look tacked on to me. But there was another way to think about Laura's wish for a porch. Colonial houses of the style we wanted to build often have large formal living rooms running from the front of the house all the way to the back on one side of the center hall. Since we would be spending much of our time in a large family room in the ell directly behind the main house, we felt we didn't need such a large formal living room.

Dividing the living room side of the main house could give us an extra

Jock Gifford's First Sketch of the Facade of Our House

room. If the living room were on the south side of the center hall, that extra room carved out of the back of the living room would have a very sunny southwest exposure. So we began to think of it as something like a "Florida room," with much more extensive windows than one usually finds in a colonial house, and with one or two glass doors opening to the garden outside the room. This room would give us much of the outdoor feeling of a porch but could be insulated effectively from both the worst cold of winter and the worst heat of summer.

Florida rooms in southern houses are often add-ons, not part of the original structure of the house. Our "outdoor room" would be integral to the main house. Jock Gifford faced the challenge of designing the oversize windows and glass doors of the room so that they would look appropriate to a house of colonial design inspiration. As Jock warmed to the concept of the room, he encouraged us to think of it as an "all-seasons room"; the name quickly took hold.

With the living room and all-seasons room designated for the space to the south of the center hall, the north half of the first floor remained for the dining room and the library. Since the kitchen was going to be in the ell behind the main house, we thought it would be best to position the library at the front of the house opposite the living room, and the dining room behind it — and thus nearer the kitchen.

On the second floor of the main house Laura and I saw need for four bedrooms and three baths: a master bedroom suite with a walk-in closet and dressing area, and a bath featuring both a large tiled shower and a separate tub; a large bedroom for Lindsey; a guest bedroom with its own full bath; a small bedroom or spare room for an overflow of grandchildren; and a full bath opening off the upstairs hall to serve Lindsey's room and the small bedroom.

In mid-November Jock delivered a floor plan for the main house. He had gotten everything from our wish list into the design but not without having to introduce some compromises no one was happy with. Upstairs, Lindsey's bedroom was not as large as we wanted it to be. The spare bedroom had translated into a room fourteen feet long, more than long enough, but only seven and a half feet wide, a little too narrow. Downstairs, the side-by-side powder room and hall closet were carving a pretty big chunk out of the all-seasons room. Between the two floors, the stairway wasn't working very well. The placement of windows we desired to give us a classic colonial facade included five windows spaced evenly across the second floor, with four windows and a center front door arranged directly beneath them. This put one of the first-floor windows right at the foot of the stairs. I could just see someone tripping on his or her way downstairs and compounding the accident by falling through the window.

A solution to the stairway safety problem was provided by Jock's wish to solve a separate problem. He had asked Laura and me whether we were "morning people," and I said I was. What concerned Jock was that the two-story main house, facing east, was going to block out morning light to the one-story kitchen and dining area sitting directly behind it. How could he divert morning light to our breakfast table?

The only way Jock could get morning light to the ell was to move the stairway so that rather than ascending from the front of the center hall toward the

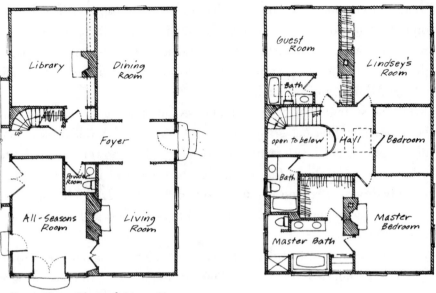

Final Floor Plan of Main House

rear of the house, it began at the back of the center hall and ascended toward the front of the house. The window placement on the front facade could then remain as we all wished because the stairway wasn't descending toward a window. By designing a shaft above the upstairs hallway up to a double skylight, Jock provided a path for morning light to flood down the shaft, over the balustrade at the top of the stairs, and on down through the opening from the downstairs hall into the ell — all the way to my breakfast! In fact, I would step into that sunlight as I left my bedroom and it would accompany me down the stairs.

This repositioning of the stairway gave Jock more flexibility upstairs, and he was able to enlarge Lindsey's room and make the spare bedroom a reasonable ten-by-twelve-foot room on the front of the house between Lindsey's room and the master bedroom. To give Jock some additional space for the all-seasons room, Laura and I agreed to eliminate the center hall closet near the front door. We would come and go usually by other exterior doors, using coat closets near them; we could live contentedly with a smaller front hall closet tucked under the staircase, sufficient for guests' coats. The space gained allowed Jock to make a major improvement in the layout of the all-seasons room.

In a final revision of the downstairs layout, we switched the positions of the formal dining room and Laura's library. Laura's room would now be tucked cozily in next to the stairway and closer to the kitchen. The dining room would be on the front of the house directly across the center hall from the formal living room. We reasoned that the formal dining room would be used only for special occasions when everyone would cheerfully tolerate the slightly greater distance between kitchen and dining room. Positioning the library on the northwest corner of the main house would enable us to sit in it and see both the shade garden we were planning to develop next to the back terrace and the rock garden on the knoll rising behind the terrace.

I was curious to see how Jock would deal with the issue of making the large windows of the all-seasons room fit into a colonial facade. He did not disappoint us. His inspiration was drawn from his first sketch for the front door. The door itself was to be solid wood, but above the door Jock had sketched in a horizontal row of small panes of glass known as a "transom light."

Today we look at such a feature as either a pleasing decorative element or as a device for letting a modest amount of light into the front hall. Originally, however, these lights had practical functions. A visitor or passerby, seeing the glow of interior candlelight through the panes above the front door, knew someone was home after dark. A passerby might also, seeing the flicker of flames through these panes, alert the household to a fire, an ever-present danger in colonial houses dependent on open-flame lighting and open-hearth heating and cooking. Some colonial builders added vertical lights beside the front door in addition to the transom lights.

Jock picked up the design motif of the transom lights above the front door and sketched them in above the two window walls in the all-seasons room. The effect was impressive. It made the window walls blend comfortably into the colonial design. There is, of course, no colonial precedent for using so much glass in a house. But I think colonists might have if the technology had been available. Pieces of glass as large as eight by ten inches were not fabricated in this country before 1792. Not until after the Civil War could a homeowner purchase a double-hung window opening at both top and bottom.

Sheets of glass of the size we were going to use in the all-seasons room were first fabricated in the twentieth century. Our "colonial" house was shaping up as an eclectic blend of colonial design with many postcolonial technological innovations.

Designing the ell began as a fairly simple exercise of providing three functions in one open twenty-two-by-thirty-six-foot space. The eastern half of the room nearest the main house would be divided into a kitchen in the northeast quadrant and a family dining area in the southeast quadrant.

Beyond the kitchen and the dining area, the entire west half of the great room would be our family room, with the west end wall dominated by a massive stone fireplace. Much of the family room would be lined with bookshelves and with cabinetry to hold — and hide when not in use — television and stereophonic equipment. Making these shelves and cabinetry of the same wood and design as the kitchen cabinets would only reinforce the sense of the entire space as one room.

The exterior walls of the all-seasons room adjacent to the east end of the ell were already designated as window walls. Jock Gifford extended this glass wall treatment along the entire south side of the ell. He ran a row of transom lights all along this wall, too, above the windows, extending his treatment of the front door and the window walls of the all-seasons room; once again, this design element helped to make a very contemporary glass wall look comfortable as part of a colonial house.

Still, when Laura looked at Jock's first sketches for the interior of the ell, she thought the glass south wall gave the great room a too-contemporary look. "We need to 'country it up,' " she said. Until then we had conceived of the great room as a one-story rectangle with a flat ceiling under a pitched, wood-shingle roof. Laura's suggestion was to install a few exposed beams in the ceiling of the ell to give the room more of a country look. Her suggestion led to a turning point in the design of the ell. Why have a flat ceiling in it at all? Why not open up the interior to the pitched roof? Before we knew it, we were dreaming of exposed timber-frame construction for the ell. It became the most dramatic section of the house.

The structural reason a timber frame made sense for the ell was that the great room had no interior walls to help bear the weight of the roof. Also, the

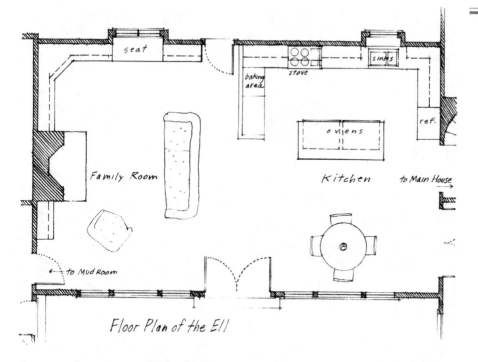

Floor Plan of the Ell

long south exterior wall would be composed largely of windows and doors, which couldn't bear any roof weight. Both the main house and the back office wing of our house would be constructed with the framing technique used most commonly in houses today called "stickframing." Exterior and interior bearing walls of stick-frame buildings have closely spaced vertical wood members called studs. The weight of the house is distributed so that each of a great many studs bears some of the weight. Interior walls that run perpendicular to the direction of the horizontal joists holding the floor above them become what is known as load-bearing walls. They are often supported by wooden beams, which in turn are supported in the basement by concrete-filled steel posts. The weight of the house comes down through many points in the interior and exterior walls until it is absorbed by the foundation and the ground beneath the foundation.

"Timberframing" uses comparatively few — but very strong — posts, beams, rafters, and other components to make a basic wood frame strong enough to bear the weight of the house without relying on either interior or exterior walls to absorb and transfer weight. The exterior walls in a timber-

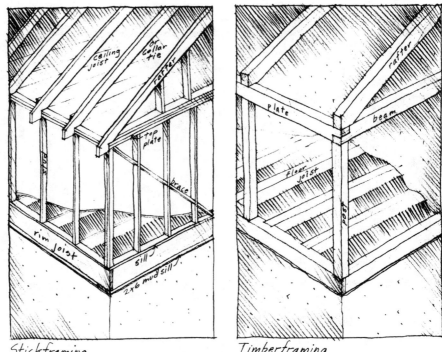

Stickframing   Timberframing

frame building, except for the timber-frame skeleton, simply protect the interior from the outside world and climate; and the interior walls divide space by function, but they don't bear the principal weight of the structure.

Historically, timberframing is an older technique than stickframing. One of the best places to see timberframing is in old barns still standing in the countryside. Regular viewers of *This Old House* invariably remember what was intended to be the renovation of a nineteenth-century Concord, Massachusetts, barn into a residence. The old timber frame turned out to have deteriorated too much from pests and exposure to weather to be saved. We pulled it down in one spectacular collapse. In a later episode, a new timber frame was raised by hand in one day by a small professional crew and a host of volunteers.

Under Jock Gifford's direction, John Murphy (an architectural draftsman who also does the final measured drawings for my *New Yankee Workshop* books) designed a system of heavy **X**-shaped trusses interspersed with other

lighter-weight rafters. The heavy trusses, also called "scissor trusses" because of the way the timbers intersect, would be eight-by-tens or eight-by-
twelves — whatever it took to meet the engineering calculations to support
the roof. The lighter intermediate rafters would have horizontal collar ties to
counter the down-out thrust of the roof. John showed these collar ties as
crossing from rafter to rafter, not down where rafter meets wall but up at the
level of the crossing point of the trusses. There was much about this design I
liked, but I thought the collar ties on the lighter rafters would impede a clear
view of the major trusses.

While John worked on the truss design, Jock Gifford was tending to a
design issue that took quite a lot of thought, but that the casual viewer would
never notice. It was a question of rooflines. The ell was a one-story section
sandwiched between a two-story main house and a back wing that had to be
at least two stories high to encompass the several functions it had to house:
double garage, utility room, office for me, bathroom, laundry room/pantry,
exercise room, and mud hall and coat closets. A major work and storage center, you might say.

Our intent was that the peak of the roof of the ell not be higher than the
eave of the main house where they met, and that the eave of the roof of the
office wing not be higher than the eave of the ell where they met. It was a
matter of having a graceful roofline. The ell would look squashed if it were
wedged between two taller sections. On a level lot it would be impossible to
accomplish this configuration. But on our property, the east-to-west (or
front-to-back) downward slope where the house was going to sit made it
possible to set the garage in the back wing several feet lower than the floor
level of the ell and main house. Jock eventually presented us with a design
we liked for the office wing that set the various rooms on four different
levels.

One level was the same as the floor level of the ell. It contained a mud hall,
two closets, and a combined pantry/laundry room. From the mud hall one-
third of a normal flight of stairs rose to the top level. It contained a large
cathedral-ceiling office for me, a full bath, an exercise room, and a hall with
an alcove large enough for a wet bar where I could keep refreshments for
office visitors.

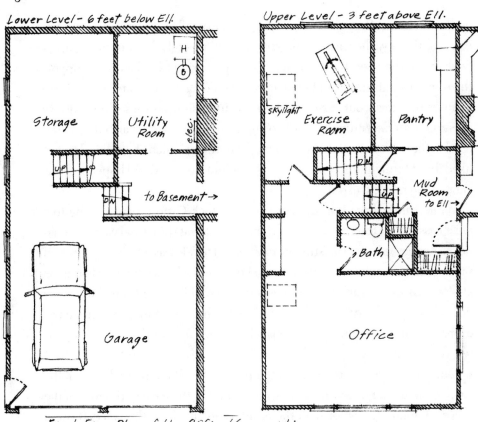

Final Floor Plan of the Office/Garage Wing

Going down rather than up from the mud hall there was two-thirds of a flight of stairs down to the garage level. In one back corner of the garage was a storage room for garden equipment. From the other back corner, steps led down three feet to the fourth level, where the utility room was positioned directly underneath the pantry/laundry room. This fourth level was the same level as the floor of the adjacent basement underneath the ell. Four levels sounds complicated, but I knew intuitively that Jock had put everything in the right place.

One of the many things I admire about Jock's work is that he isn't possessive about the design process. There were a few instances in which he suggested that I work out the details myself because he felt it was within my capacity to do so. When I told him that I planned to have the timber frame for the ell cut and assembled at Tedd Benson's company, he immediately suggested that I involve Tedd in the design aspect of the frame. Tedd is the timber-framer whose fine craftsmen designed and managed the installation of the timber-frame barn/house we televised in Concord. In my opinion he is the premier timber-framer in America. He and others have stimulated a resurgence of timber-frame houses all across the country.

"Nothing's close to Alstead," Tedd warned me as he gave me some directions by telephone to Alstead Center in New Hampshire, where his workshop is located. He had invited Laura and me to visit his shop in early December of 1991 and to bring along John Murphy's sketches so we could discuss further design ideas and see wood samples. Alstead was country, all right. The two-lane road wound past a small common with a white-steepled church in a landscape of tin-roofed barns, pastures dotted with cows and horses, and a sprinkling of trailer homes.

Tedd's company occupies an unpretentious barn and ell set against a hillside, much the way our house would nestle against the granite knoll. In the shop, two work spaces meet at a right angle. In one of them the roof is supported by trusses made of different kinds of wood — Sitka spruce, Port Orford cedar, red oak, white oak, pine, and fir. Shavings and sawdust cover the floor. A huge wood-burning stove warms the shop in winter.

The workshop crew, most of them wearing T-shirts or sweatshirts declaring themselves members of the "Beam Team," work in part with traditional hand tools of the craft. Carefully sharpened and oiled chisels are the instruments for some of the shaping and fitting. However, there is also plenty of evidence of their use of high-tech, state-of-the-art power tools.

Tedd led us out back of the workshop to look at wood in his drying sheds. The twenty-five-foot-tall sheds held timbers twenty, thirty, even forty feet long. (The term "timber" in construction refers to lumber of at least five inches in its smallest dimension.) Some were green (recently cut), but many more were weathered to gray and studded with rusting bolts: timbers salvaged from abandoned factories and mills. There were pieces that weighed a ton or more each. I ran my finger over the end of a huge old timber. In the varying widths of its growth circles — some narrow, some wider — I could see the history of the climate in which the tree once grew, and make a guess of its age.

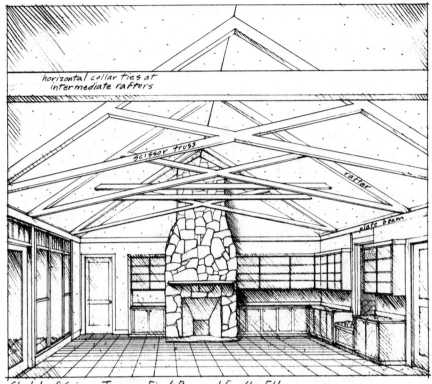

horizontal collar ties at
intermediate rafters

scissor truss

rafter

plate beam

Sketch of Scissor Trusses First Proposed for the Ell

Judging from the markings on several Douglas fir timbers reserved for a house in Colorado, Tedd's almost twenty-year-old business now takes him far from tiny Alstead. I noted that his outdoor power tools were simple despite the size of the timbers being handled. A once portable, now permanently set into the dirt band-saw mill operated by a single person handles all the preliminary milling before timbers are brought into the workshop.

Each of the trusses we needed to bear and distribute the weight of the roof in our twenty-two-foot-wide ell would be basically triangular in shape. From the ridge beam at the peak of the roof, a heavy timber rafter would come down each side of the pitched roof to rest on a timber post set into the side wall. Where the rafters met the side posts, a third beam would span horizontally from one wall to the other to complete the triangle.

As the weight of the roof bears down on the supporting rafters, their impulse is to push the walls of the house outward. The function of the horizontal timber is not so much to bear weight as to check the outward thrust of the roof weight. The tops of the walls can't be forced or pushed outward because they are joined by the horizontal timbers.

In thinking about the timber for our trusses, I had already eliminated two softwoods: pine (its knots are too regular in appearance and spacing) and Douglas fir (too red). My first choice was a hardwood, white oak. Hardwood construction lumber is usually more expensive than softwood. The leafy deciduous trees from which hardwood lumber is milled each require more growing space than most softwood conifers; they grow more slowly and often not entirely straight. The softwoods grow straighter, faster, and taller, as a rule.

Most construction lumber is softwood. When it's first cut, as much as 50 percent of its content is water. More than 25 percent moisture content creates an opportunity for dry rot to set in, so all construction lumber — soft or hard — must be dried before using. Drying has the additional benefit of stabilizing the wood and minimizing its capacity to warp or crack. Light boards, not thicker than two inches, will air-dry quickly to less than 25 percent moisture content. Timbers of the size we needed would be air-dried for several months and then finished off with kiln-drying to get down to the 15 percent moisture level that would be best for making our trusses.

Knowing my preference for white oak, Tedd showed us some mixed white and red oak timbers the Beam Team was working on in his yard. "The available trees are so small," he said, "that you don't get a good-sized timber. It's not old growth, and the wood isn't as stable. I think you'll get a lot of checking [cracking] from these timbers." There went my dream of white oak. Tedd recommended Port Orford cedar from the Pacific Northwest for its stability — it wouldn't shrink much as it dried or expand much in humid weather; it would have relatively few twists, bows, checks, or splits. But Port Orford cedar is so clear (free of knots) that it differs little in appearance from Sitka spruce, which has most of the same virtues and one more: being in greater supply, Sitka spruce costs only about half as much as Port Orford cedar.

Whatever the wood selection, we were getting into an expensive project with the timber-frame ell. It was time to be practical. We decided on the spruce. Later, we decided to use Port Orford cedar just for the posts and plate beams because they are the only parts of the timber frame that will be subjected to both the cold of winter on the outside and the heated space inside without protection from insulation or a vapor barrier. This could cause condensation to form on the outside face of the posts and beams, making them rot. Because of Port Orford cedar's resistance to rot, this problem would be minimized or eliminated.

Laura and I visited the design studio upstairs in the barn, where architects and engineers work at computer stations, then went back down to a conference room at the shop level with Tedd and his chief designer, Bill Holtz. As we looked at sketches of the ell, I described the setting. "This is our main living area," I said, "where we want the indoors and outdoors to work together. From the French doors in the center of the windowed south wall you'll step down two steps to a terrace of granite slabs, then from the terrace down another few steps to a garden. The driveway is beyond the garden. On the other side of the driveway it's all forest, fifty acres of it. Very private. So the great room of the ell is where we'll live, hang out, entertain. The trusses should make a statement that this is *the* room."

"Will visitors enter through these French doors?" Tedd asked. "They could," I said, "but most visitors will drive into the guest parking area at the front of the house and come to the front door without yet having seen the ell behind the main house. Coming in the front door they will immediately see through the center hall into the ell, a long, unexpected, dramatic view that ends at the massive stone fireplace. They'll come in, and say, 'Aaah . . .' " Because of the long view, I added, I wanted to minimize the number of trusses so that each one could be that much better seen. If there was a truss every four feet, I believed, each would obscure the view of others.

Tedd and Bill thought that four trusses along the thirty-six feet of the ell would be sufficient. There would be a heavy timber beam along the ridge of the ceiling between the trusses. There would be a heavy timber plate beam at the top of each side wall, sitting on top of four heavy posts on the south wall where all the windows were, and on top of a two-by-six stud wall on the north side. Because the latter side is a northern exposure, there were going to be only two windows — one in the kitchen and one in the family room. To allow thicker insulation in the north wall, that wall would be stick-framed in two-by-six studs. Between the trusses there would be lighter rafters going from the ridge beam down to the plate beams. These lighter rafters would also bear some of the roof weight.

Tedd suggested that the horizontal timber (he called it a "bottom chord") joining the two rafter timbers of each truss be slightly arched and made of laminated wood rather than being a solid member. The Beam Team, he said, could run a band saw through each bottom-chord timber, cutting it into one-inch-thick strips. After the strips were planed, they would be roller-painted with an epoxy adhesive so transparent when dry as to be undetectable to the eye; then the strips, still wet with adhesive, would be clamped together against a large curved jig to form the desired arch. When complete, the arched laminated bottom chord would be as strong as a straight, solid wood beam.

Under the most stressful conditions — a heavy accumulation of winter snow and ice on the ell roof, let's say — the roof would be exerting a lot of downward force on the rafters of the truss, which would result in the walls being forced out and the arched collar ties flattened. To counter that force, Tedd added a king post to each truss. The king post rises from the center of the bottom chord vertically to the ridge beam, bisecting the triangular truss into two smaller triangles. Its function is not to bear weight but rather to prevent the arched bottom chord from flattening out under a heavy load — to hold it *up*.

At the ridge, the king post would be attached to the ridge beam with tra-ditional mortise-and-tenon joinery. There was plenty of wood at the ridge to make a strong joint, but where the king post met the bottom chord, Tedd and his engineer thought that a traditional timber-frame joint would not do the job. A metal brace would be required — something a timber-framer does not like to give in to. Rather than using ordinary plate steel for the brace, Tedd recommended we consult a metalsmith he knew to explore something a bit more artistic yet functional.

Now that we had a structure for the roof, how would we close it in? If we had designed a flat ceiling in the ell with an attic space between the ceiling and the pitched roof above it, we could then have laid conventional batt insulation in the attic space. Opening up the ell ceiling prevented that. There was no attic space left. How would we insulate the roof to meet the requirements of the

Massachusetts building code? I could use dimension lumber (lumber cut to size and stocked for the building industry, usually two to five inches thick and five to twelve inches wide) and frame a space on top of the timber frame that could be insulated and vented, but that space would have to be thicker than I wanted. If, for example, I used two-by-tens to frame the space, it would push the ridge line high enough to put it above the eave line of the main house. That was unacceptable. I definitely wanted the ridge of the ell roof to tuck in under the eave of the main house, where they met.

Tedd Benson already had an answer to this dilemma. His company could provide — through a subcontracting arrangement — the same kind of stress skin panel I had first seen used on the Concord barn/house and for the same purpose: applying insulation to a timber-frame ceiling without obscuring the frame. "Stress skin panels" are three-layer sandwiches in which the outer layers of wood or wallboard enclose and protect a center core of rigid insulation. In our case, the inside layer would be plasterboard. The insulation core or center layer would be a three-and-a-half-inch-layer of polyisocyanurate with a high insulating value. Between the rigid insulation and the roof would be a half-inch outer layer of oriented strandboard made from pieces of wood fiber compressed and bonded with phenolic resin. A very impressive sandwich!

In sequence, as Tedd outlined the process to Laura and me, the timber frame would be installed, the stress skin panels applied to the outside of the frame, and then a roofing system applied to the outside of the stress skin panels. I would need to make sure that the roofing system allowed for venting between the stress skin panels and the shingles in order to prevent the rotting potential I had noted in the roof of the house we didn't buy, where shingles had been applied directly to the plywood subroofing. Benson's design would keep us efficiently warm or cool as the season dictated, but not without a high start-up cost. The timber frame, installed, with the panels but not the roofing, would come to about $30,000. When we added the cost of roofing, windows and doors, flooring, masonry, cabinetry, plumbing and electrical work, lighting fixtures, and kitchen appliances, we were on the verge of ordering a $100,000 room. We had "countried it up," all right, but maybe at city slicker prices.

On the drive home from Alstead, Laura and I reviewed what had been accomplished in the design of the house. The dream was taking shape. Our conversation turned to fireplaces. There were going to be four fireplaces in our house: in Laura's library, the family room, the formal living room, and the master bedroom. I planned to put Rumford fireplaces in all of them. What Laura admired in a Rumford fireplace for its appearance was also a smart choice for its efficiency.

Count Rumford, an American-born inventor (also a scientist, landscape architect, and statesman), had talent reminiscent of his contemporaries Jefferson and Franklin. During a visit to London in the 1790s, Rumford began to think about a solution to the smoking fireplaces that vexed many Londoners. To inhibit downdrafts from blowing smoke out into the room, Rumford narrowed the throat where the firebox opens into the chimney flue. He reduced the overall size of the firebox, slanted its side walls, and made it shallow rather than deep. The fireplace therefore radiated its heat into the room more effectively, and drew less heat up the chimney.

Even today, some disappointed owners of new houses find that poorly engineered or built fireplaces smoke up the house. It's difficult and very expensive to tear out a fireplace to try to correct a smoking problem. More frequently, however, our concern today is with heat conservation. The blazing hearth may be drawing precious heat out of nearby rooms and evacuating it up the chimney. In each room where we planned to have a fireplace, but especially in the family room with its soaring ceiling and great space to heat, we needed to have an efficient fireplace.

Before we could relax by a cheerful fire in a Rumford fireplace, however, we still had a house to build. It was getting close to two years since we had first talked about moving and had begun to look at houses. Now the winter of 1991–92 was upon us. Little could be done on the site before early spring of 1992. The yearly January-to-March pilgrimage of *This Old House* from Boston to a climate where construction was easier for filming was about to take me back and forth across the Atlantic to London for much of several weeks. But I was getting very itchy to move on from architectural plans to some hands-on labor. I yearned to see a bulldozer on the property.

Wetlands

WETLANDS, you could say, were the reason for the very existence of the town into which Laura, Lindsey, and I intended to move. According to one local history, this rural area was almost routinely isolated from the nearest town in eighteenth-century springtimes by "fluds." What irked the locals most was that they had to pay church taxes but received only sporadic clergy attention during the long flud-and-mud seasons. A sympathetic colonial legislature responded to a petition and gave our area its own congregational meetinghouse in 1754; by 1805 the area had become an independent town.

A mark of the prevailing geological wetness is that 1 percent of the total acreage of the town even today is cultivated as cranberry bogs. Despite a town ordinance requiring that every building plot be at least two acres in size, finding a piece of undeveloped land in the town that doesn't require a variance related to wetlands is almost impossible, I learned.

A local Conservation Commission — appointed by elected town officials, but unpaid — is responsible for local application of the state wetlands protection legislation. It is now virtually impossible to build directly on wetlands — on a tidal flat, for example, or on the bank of a stream or in a freshwater swamp. The law further requires a variance granted by the Conservation Commission before construction of any building within a buffer zone of one hundred feet next to wetland.

I had been concerned ever since Laura and I first explored the property in January of 1991 that the small trickle bubbling out of the ground near the western boundary made the area, though not marked as such on the subdivision plan, wetland. But a year later I still had not done anything to clarify the situation; chalk that up, I guess, to my natural reluctance to take on local governmental bureaucracy. Joe March, a civil engineer from the firm that was rendering a site plan and redesigning our septic system, reinforced my own suspicion. "That water over there," he said to me, pointing to the western boundary, "you'd better have it marked and flagged." He cautioned me that if I began construction and then someone from the town noticed there was wetland during a routine inspection, I could be held up indefinitely and might have to make expensive modification of my plans.

My intention was to site the house so that there would be enough space between the house and the western boundary to build a detached workshop approximately thirty by forty feet, just a bit larger than the studio workshop I use for *The New Yankee Workshop.* In the site plan Tom Wirth prepared for Laura and me, the western edge of the workshop was within ten feet of where water trickled out of the ground! I knew that wouldn't pass review by the Conservation Commission if they judged the trickle and its vicinity to be wetland. Shift the house and workshop eastward, I told Joe. But there was no way we could shift them eastward enough to clear the one-hundred-foot buffer zone entirely and still keep the house on the slope behind the knoll.

Feeling a little anxious, I applied to the Conservation Commission for a variance permitting us to build a workshop wholly within the buffer zone (and yet not too close to the trickle), and a house with just one corner within the buffer zone. If, as seemed likely, the commission identified the trickle as wetland, no building permit could be issued until I was either granted a variance or relocated the buildings entirely outside the buffer zone. Laura's and my petition became an agenda item for March 5, 1992. A notice of our petition was posted in the local paper. Laura and I were required to notify everyone — in writing — with property abutting ours of the hearing. (A couple of those homeowners did attend the hearing.)

The hearing was rather informal. I found the Conservation Commissioners to be an impressive group; they included a botanist, a professional hy-

Aerial view of the property after preliminary clearing of the building site. The access road to four properties, including ours, can be seen at top left of the photograph; the silt barrier to protect wetland is just visible as a curving heavy line at lower right.

drologist, and two teachers. To accommodate their limited availability, the town had appropriated funds for them to hire a staff person, Pat Loring. She visited sites before hearings. Unless she spotted something unusual that the commissioners thought warranted their personal inspection, the commission generally accepted her recommendations.

Pat Loring was as confused about the location and boundaries of our property on her first visit as I was on my first visit, but she eventually found the site and walked through. She noted the granite knoll and the trickle near the western boundary, which she was sure constituted wetland. It was immediately clear to her that I didn't have many options for a good building site on the property. "The town would love to move all construction outside the one-hundred-foot buffer zone," she told me later, "but that seldom works out. As long as the work isn't in a wetlands resource area, there's no reason to deny a variance."

I gave a lot of thought to how best to relate to the hearing before the commission. Would my public television exposure be a benefit or a hindrance? Would they go out of their way to be tough just so that no one could accuse them of being easy on a "celebrity"? Should Laura and I represent ourselves or should we let someone like our lawyer represent us? Eventually we decided it was best to represent ourselves personally.

The hearing began with general discussion of the plot plan that had been prepared by Joe March. What concerned the commission most was that the workshop building on the plot plan I gave them was described as a "barn." I don't remember who first referred to the proposed shop as a barn — maybe *I* did — but I think the point was simply that the shop would look more like a small barn than any other type of shed.

The commission jumped to the conclusion that I was going to have farm animals — horses, most likely, as is common in the area, but maybe cows or even pigs. They had instant fears about manure and urine leaching into the wetland. A secondary concern was that I might use pressure-treated lumber in construction of the barn or of a paddock. They believed pressure-treated lumber releases toxic chemicals that could leach into any adjacent wetland. My information about pressure-treated lumber indicates this is not the case, but I guess they didn't want to take any chances. I didn't see anything to be gained by arguing a point that wasn't relevant to my plans anyhow. What I did emphasize, quickly and unequivocally, was that my intended detached workshop was not going to be a barn for animals.

At first, the commissioners proposed that we solve the wetland problem by simply moving the buildings enough to the east to get them out of both wetland and buffer zone. But we knew the site better than they did. We had spent a lot of time and money siting the house and workshop, and there wasn't much room to maneuver and still have a desirable placement of the house.

The commissioners wanted to know how much water was flowing and where it originated. As we were discussing the contour of the land near the trickle, I mentioned that while my property sloped fairly steeply down to the area of the trickle, the grade on my side wasn't as steep as the grade coming down to it from my neighbor's property on the far side of the trickle. There the terrain dropped off so sharply that it didn't seem natural. "I don't know what's going on there," I pondered. "The way the grade drops off, it looks like fill, maybe ten feet from the trickle." I was actually setting up a little di-

The terrain on which we were setting our house sloped both north to south (background to foreground) and east to west (right to left). I wanted to minimize how much the south end of the main house loomed above grade. My first preference for the house's elevation left 3½ to 4½ feet of exposed foundation along the south end and would require the north end to be excavated about 3 to 3½ feet.

Subsurface rock encountered in excavating the basement caused me to raise the house 2 feet. Now the exposed south foundation ranged from 5½ to 6 feet and the excavation at the north end would be reduced to 1½ feet.

After Herb Brockert regraded the site, the house nestled snugly there with only 8 to 12 inches of exposed foundation.

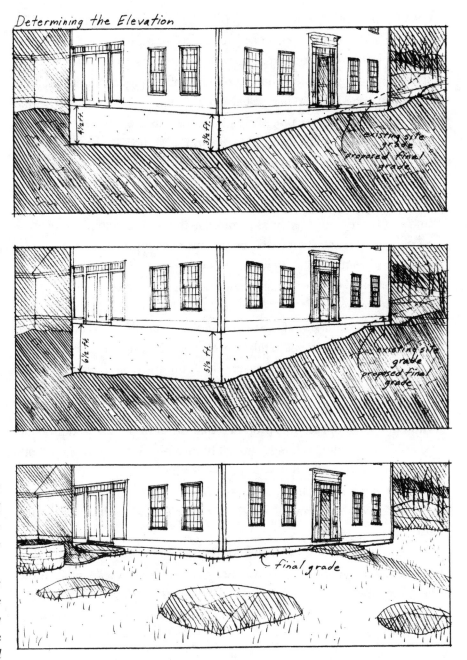

Determining the Elevation

version. If my observation was accurate, the contractor who built my neigh-bor's house to the west had taken the easy way out. Instead of hauling away unwanted material from clearing the site — and maybe other construction debris as well — he had dumped it over the bank at the back of the property next to what had become now our property line, and then covered it with enough soil to hide it.

The hearing on our petition concluded with the commissioners accepting Pat Loring's recommendation that they should all visit the site to see the unusual configuration of the knoll and the stream. They all expressed igno-rance of what might have occurred on my neighbor's bank but said they'd take a look during their visit the following Sunday.

I welcomed a visit of the commission to our property. Although I had to fly to Chicago for a personal appearance on Saturday, I'd be home on Saturday night in plenty of time for the Sunday visitation. But again, and not for the last time, weather was going to intrude on our plans. Laura was getting ready for bed in Massachusetts on Saturday night when I called from O'Hare Air-port in Chicago. A late winter storm had left all planes indefinitely grounded by near blizzard conditions. "It's bad here," I said. "I don't think I can get home in time. You'll have to lead the commission through the property."

"Norm," Laura said, "you *have* to get back."

"I'll do everything I can," I answered, "but don't count on it."

As I recall, I slept soundly that night, but Laura said she scarcely slept at all. The next morning I was still in Chicago. Laura, armed with a bag of fresh bagels and cups of hot coffee, waited nervously at the site. Tom Wirth had marked the approximate location of the house and workshop with small stakes and plastic ribbons for the benefit of anyone who wanted to see where the house and workshop would sit. The five commissioners arrived about fifteen minutes behind schedule. They were dressed prudently for the day in heavy boots. Laura had worn only sneakers. Dutifully, she led them into the wooded lot. They wanted to see everything, she said, and they would huddle aside from her and talk to each other while her toes gradually went numb from the cold.

Then one of the commissioners spotted the neighbor's steep bank on the far side of the wetland. They had no doubt, Laura reported, of what had happened. "Stump dump!" they growled to each other. "Everything changed at that moment," Laura said. "I knew we were fine. Our neighbor had a much bigger problem."

We went to a second hearing on March 26. After very little discussion, the Conservation Commission approved our petition for a variance. The variance required that we construct a silt barrier composed of hay bales and plastic sheeting between the construction area and the wetland to prevent any contaminants from leaching down into the wetland during construction. We could not modify our proposed grades within the buffer zone without further approval. During construction, Pat Loring would be monitoring the site regularly for the commission.

The first of several permits — this one from the Massachusetts Department of Environmental Protection — was almost ours. We had only to observe a thirty-day waiting period, during which any townspeople who wished to could appeal the commission's published action on our petition (none did), and then we had to file the permit with our deed, which I did myself at the courthouse in Lowell. Outside the town hall after the commission had said yes, in a spring rain, Laura and Lindsey and I jumped and hollered with glee. We could soon begin to clear our land, which had — the local weekly newspaper conceded in its report on the hearing — "looked unsuited to animals."

As soon as we received the environmental permit certificate, I built a small wooden box, which would both display, as required by state law, and protect it. I attached the box to an oak tree that I knew would not be cleared for construction. I also made a sign with our permit number and displayed it on a tree at the driveway entrance.

When Joe March prompted me to start dealing with the wetland issue, he was working on a revision of the septic system plan. I was confident the Board of Health would quickly approve Joe's new design. Everything had changed from the plan approved for the previous owner's unbuilt house except the specifications for the septic field. I had vowed not to change the field in order to avoid having to retest for soil porosity and depth of water table. (Just to indicate the unpredictability of the building process, we would over time file three different plans with the board as the septic system evolved; though I

didn't ever change the location of the septic field, I did eventually lower it, necessitating some of the very retesting I had vowed to avoid!)

When we sited the house, we created an elevation problem between the septic tank and the septic field. The septic tank was going to be buried in front of the house. There was no problem with the waste removal system working by gravity flow between the house and the tank. The problem lay between the tank and the leaching field. They were so close to the same elevation that gravity wouldn't guarantee a natural downhill flow from tank to field.

If we ran a straight line from the tank to the field, the line would be seventy-five feet long. But that straight line ran straight through the knoll. I believed the knoll was mostly solid rock — too expensive to blast through. I began to resign myself to a more complex septic system than I preferred. The plastic pipe from tank to field would have to take the long route around the knoll and then along the edge of the driveway. The waste fluids would have to be pushed through the line with the assistance of a mechanical pump.

The silt barrier of posts, plastic sheeting, and hay bales at ground level. Hardly glamorous, but satisfactory to the Conservation Commission to keep debris from our construction out of the wetland.

Joe March's plan for getting waste from the house to the septic field involved two separate tanks. All sewage would flow into the main septic tank, where solid waste would remain. Liquid waste would drain into an adjacent underground pumping tank. When liquid waste reached a set level in the pumping tank, the pump would engage and force the liquid out to the trenches in the septic field.

Plenty of things could go wrong in such a system, of which the worst is pump failure. I decided to have a backup pump in the pumping tank and an alarm system to alert me to any failure of the main pump. The backup pump would engage automatically as needed, but the alarm would prompt me to have the main pump fixed while the backup pump kept the system operating. My father once had both a sump pump and a backup pump in his basement to remove any storm water. But he didn't have an alarm to signal him on the one occasion when both pumps failed. What a mess he had on his hands!

I calculated what it had cost me just to get this far — not a stone moved yet, not a tree felled, not a rough driveway bulldozed, not a shovel of earth lifted. The engineer's work on the septic system redesign, the engineer's re-siting of the house to move it back from the wetland, the survey of the wetland required by the state, and the study of the wetland by a botanist to determine by vegetation where the wetland ended — all of this had cost about $2,000. A small trickle not even large enough to support the smallest tadpole had already cost me $2,000 in out-of-pocket expense plus the expenditure of much time negotiating my way through the process of getting approvals; all I had to show for the expense were some strips of colored plastic tied to branches and bushes to show where I *couldn't* build.

"Fire in the Hole!"

THROUGHOUT THE WINTER of 1991–92, Laura and I discussed the new house almost every evening I was home — that is, when I wasn't in London video-taping programs or on the road making personal ap-pearances. We focused mainly on questions related to finishes — floors, cabinetry, counters, and so on. Laura deferred to me on construction ques-tions. Wherever I was that winter, I lay in bed at night thinking about the options. There were scores of issues to think about. Super insulation or merely adequate insulation? Stain or paint on the exterior clapboards? Ra-diant under-floor heating or conventional baseboard units? Every room in the house had its questions to answer.

During those winter nights I also began to take a mental inventory of the elements in the construction of the house that I would reserve for myself. I would definitely build the stairway in the main house with its winding treads and its curved balustrade at the top, I also reserved for myself the installation of interior finish woodwork, such as moldings, mantels, and wainscotting (but not necessarily before we moved in, though I didn't mention this to Laura yet). It was debatable whether I would have time to build the cabin-etry and shelving for the kitchen and family room, but I intended at the very least to install them. Maybe I could install the exterior clapboards and trim on the house and paint or stain them. Whenever possible I would definitely become an extra member of whatever crew was working on the house, lending a hand as needed to push the work along.

Oh, yes, I'd also be the janitor, cleaning up after a day's work, and also reminding the various tradesmen also to clean up after themselves. I like an orderly site. One of my little dreams for the construction process was not ever to have a large trash container on the site. Those ugly oversize wastebaskets are meant to encourage neatness, but in my experience they encourage a kind of casualness about maintaining the site.

Overall, I was wearing three different hats in this enterprise. I was co-owner with Laura. I was going to be general contractor. As general contractor, I planned to subcontract certain aspects of the work out to myself. Here's a bit of advice I'd give to almost anyone outside the field of construction who is planning to build a dream house: don't try to be your own general contractor. Since I enjoyed the benefit of having been a general contractor for several years, I planned to ignore this wisdom, but not without considering the pitfalls. Never as a general contractor did I try to build a house and hold down another full-time job at the same time. I'd have to be careful. The perspectives of owner-dreamer and contractor-builder are so different in terms of purpose, motivation, and emotional investment that I had to be sure that I didn't confuse them as I undertook both roles at the same time.

The advantages to an owner of being his own general contractor are two-fold. He has the pleasure of being closer to the work as decisions are made and the house takes shape; and he saves the portion of the total cost — perhaps 20 percent in a typical situation — that flows to the general contractor for his work and profit. The general contractor looks at a house as a businessman does. In the case of a single house like ours, he probably does very little of the work with his own staff. He subcontracts pieces of the job to various trades and specialized businesses: framers, electricians, plumbers, roofers, painters, and so on. He becomes the conductor who keeps a small orchestra of different players working in sync to build the house well and on a predetermined schedule — and profitably.

Each of the subcontractors in turn is an independent business with its own clients and schedules. In an ideal world, the subcontractors would have uniform demands on their time, but, as you know, the world doesn't work that way. On the day when I desperately want the electrician to be working on *my* house, he may be responding to the earnest supplications of another of his clients. Yet it is in the very nature of house building that everything has to be done in the right order, and when proper order is violated, there usually are problems and extra costs.

Most owners don't have enough flexibility to be on call at all times during the workweek, or to adjust to weather delays, missed deliveries, and subcontractors who don't appear as promised. My schedule is demanding, to say the least. So I made the compromise with myself of deemphasizing the necessity of building on a tight schedule. It's just as well that I made such a concession at the outset. My dream was to move into a new house about a year after we acquired land for it. On the first anniversary of the closing on the property, our excavator was still trying to decide where to drill holes for the dynamite to blast hidden rock ledges where we meant the foundation of the house to sit.

On February 7, 1992, Tom Wirth sent us a drawing showing a revised siting of the house to take into account the shifting of the house eastward to keep it almost entirely out of the wetland buffer zone. Once that siting was final, work to make a visible difference to the property could begin. I waited until early April. Then I just couldn't wait any longer. A spring snowstorm was still a possibility — in New England maybe even a probability — but there were enough signs of spring to overcome any caution reason might impose. We didn't have our building permit yet, but I believed I didn't need one before clearing trees and brush, building a rough driveway, and even excavating for the basement and foundation.

I called in Roger Cook, a landscape contractor, as the first subcontractor to work on our property. Until Roger and a tree-removal crew he hired (I guess that constituted a subcontractor hiring a sub-subcontractor) cleared out some trees and underbrush, no one could get equipment onto the site to do anything else.

Roger was one of the crew of the landscape contractors who handled the first two projects televised on *This Old House*. Even then, I knew that sooner or later he'd branch out on his own. He always kept the job moving. He is as levelheaded and good-natured as he is rugged; I've never seen him lose his

*March 1992. The
building site has been
cleared by Roger
Cook. Two of the
stakes set by a
surveyor to mark
boundaries of the
house can be seen on
the right side of the
clearing. Note how the
site slopes down both
toward the camera
(southward) and to
the left (westward).*

temper. On the last job I managed as a general contractor, I asked Roger if he
would do some demolition for me. It was winter, and there wasn't much
landscaping work to be had and very little snow to plow. He took the job and
did it as well and quickly as I had counted on.

Before dropping any trees on our property, Roger's crew carried in the
straw bales and heavy plastic sheeting to build the silt barrier between the
construction area and the wetland. Stakes they pounded into the ground
along the line of the barrier guaranteed that the straw and sheeting would
not slide downhill toward the wetland. The barrier didn't look very attractive, but it didn't need to; once the house and workshop were constructed
and the surrounding soil stabilized with plant growth, the Conservation
Commission would authorize its removal.

A surveyor from the engineering firm that redesigned our septic system
had already staked out the perimeters of the house and workshop according
to the revised site plan they prepared and that was approved by the Conservation Commission. Everything within, or even close to, those markers had

to be cleared. Roger and I, working with the plan, marked where the driveway would come in from the access road so that trees and brush could be removed there as well. Elsewhere, in the general area where the buildings and surrounding garden would stand, we marked trees selectively for cutting or sparing — taking into account their maturity and variety. Fortunately, there were only about twenty large trees to take down, but there were a lot of small trees and brush of little value to anyone. On the whole, we wanted to remove as little as possible.

For several days, the racket of chain saws must have made our nearest neighbors aware that the first phase of land clearing had begun. The crew sawed all of the hardwood into two-foot lengths and set it aside in a huge pile to season, eventually to be split for burning in our fireplaces. We wouldn't need to buy firewood for a few years. We might even sell some of it and use the proceeds to pay for part of the landscaping.

All the brush and smaller branches were fed through a chipper to make a small mountain of mulch available for a later stage of landscaping. I didn't keep any of the softwood for our fireplaces. It burns more quickly and not as hotly as hardwood, and the softwood resins in the smoke collect on the sides of the flue as the rising smoke cools, creating a potential for flue fires.

When Roger and his crew left, the building site and the path for the driveway were clean and neat except for tree stumps. At this point the surveyor came back and staked out the house in a way that would enable the excavator to dig the foundation accurately. Neon orange stakes pounded in by the surveyor already marked the corners of the main house, ell, and office wing. These stakes would quickly be obliterated when the excavator began to dig. The surveyor therefore added some stakes called "off-sets" about twenty to thirty feet away from the foundation. Two of the off-sets were up on the knoll; one was lined up with the facade of the main house, the other with the west wall of the office wing. Other off-sets were staked out from the south side of the house. These stakes could be checked at any time to make sure that the foundation was being correctly positioned.

On a cool, overcast morning in mid-April, Tom Wirth, Herb Brockert, and I stood by Herb's truck looking at a plot plan Herb had spread out across the hood. Herb was my excavator, responsible for constructing the driveway,

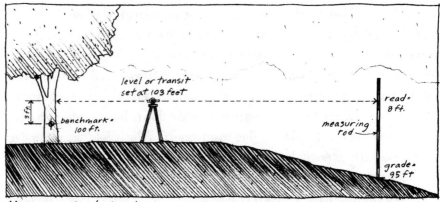

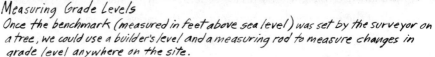

Measuring Grade Levels

Once the benchmark (measured in feet above sea level) was set by the surveyor on a tree, we could use a builder's level and a measuring rod to measure changes in grade level anywhere on the site.

excavating the foundation, installing the septic system, and, no doubt, a few other tasks requiring his equipment and skill.

The three of us were discussing two issues that became apparent once the site was cleared. The first issue concerned how the house would fit into the slope. When the slope had been wooded, its north-south decline away from the knoll was obvious even to a casual observer. Now that the area was cleared, we could see that the site also sloped down from east (facade of main house) to west (west side of office/garage wing) more than the plot plan indicated.

The surveyors had left a benchmark on a tree about fifty feet east of the main house location when they worked out the site plan. Using two tools — a builder's level (which I later replaced with a laser level) and a measuring rod — I could calculate grade levels and potential foundation heights or elevations myself based on the benchmark. All I needed was a second person to hold the rod.

Grades and elevations were expressed in numbers representing feet above median sea level. The actual numbers weren't important; it was their relation to each other that meant something. The top of the knoll, for example, was about 250 feet above sea level; the slope where the house would sit near the base of the knoll was between 241 and 229 feet depending on the point being measured. The long north wall of the house parallel to the knoll didn't present any elevation problems; it dropped less than a foot from east to west.

Here are some of the grade changes that concerned me. The grade level along the front of the main house dropped almost five feet. Along the back of the main house it dropped almost seven feet. From the highest point, the northwest corner of the main house, to the lowest, the southwest corner of the garage, the drop was eleven and a half feet.

These substantial differences in the natural grade levels of the site within the footprint of the house constituted a challenge in a house that wasn't meant to look eccentrically multilevel, but rather to look as though it had been built on a level lot. To some extent we could modify the slope in either direction by moving earth around, filling in, but the question was how much of that we could and should do. One thing I wanted to avoid was having the sides of the house on the downward slopes loom up out of the ground in such a way as to expose a considerable amount of the foundation. That wouldn't look good in any house because of the exposed foundation, but especially not in the case of a colonial-style house meant to hug its site in a very snug way.

I had been thinking about establishing the elevation of the finished floor of the first story of the main house and ell at 238.5 feet, over two and one-half feet below the natural grade level of the northwest corner. If I raised the house above 238.5 feet, I would, because of the natural slope, bring the southeast corner well over three feet out of the ground.

Tom Wirth, thinking about the existing beauty of the site, wanted to disturb the natural slope as little as possible. He advised keeping the main house floor at 238.5 feet even though the north wall of the main house would then be a couple of feet below grade. In Tom's view, the best way to deal with the north end of the main house being partly below grade was to make the north wall a retaining wall of poured concrete instead of an above-grade exterior wall sheathed in clapboards.

After thinking about Tom's suggestion, I decided that I didn't want the north wall of the main house to be a retaining wall, some of it below grade. I wanted the house to seem to be snuggled down in its setting, but I didn't want it to look as though it was already sinking, if slowly, into the ground. If necessary, we would raise the elevation of the floor of the first story a foot, from 238.5 to 239.5, and excavate enough of the earth around the highest northwest corner of the house (where the natural elevation was 241.2) that all of the first floor would be above grade. I also decided that I wanted to have a minimum of six to eight feet between the north wall of the main house and the knoll. The house should be close to the knoll but also clearly separated from it. I knew it would be a challenge, but I felt that we could raise the grade along the south wall of the main house in a way that would not look like we built a deliberate embankment to make the house appear to hug the ground.

Herb Brockert was prepared by virtue of his trade as excavator to rearrange the slope with fill and grading to get whatever final effect we wanted. But his bottom-line advice was that we didn't really know what we were getting into in a site so obviously characterized by subsurface rock, and so we'd simply have to figure things out as we went along — very wise counsel indeed, as it turned out.

I think of Herb's career as reflecting many of the pleasures and most of the headaches of being a small operator in the construction industry. I'm naturally sympathetic because I was a small operator, too. Herb had attempted twice with the same partner to launch an excavating business. When the first attempt failed, he went to work for a time with a huge general contracting firm. The second attempt ended in disagreement (as my partnership once ended, though not in disagreement). The two partners divvied up the machines, and Herb gave himself five years to get established on his own. It was the 1980s; everyone was making it. Herb worked up to ownership of eight machines, each a very large investment. He had five men working for him, building roads and tennis courts, dredging ponds.

Then in 1990 the economy fell apart. It happened within a month, Herb says. Once there was work; then, suddenly, there wasn't. He laid off all his men and went back to residential excavation. He's become a familiar, relaxed figure on *This Old House*, but Herb remembers the way his heart thumped the first day he was on camera; he thought the mike might pick up the thump. Work in general was still slow for him when I asked him to be excavator for our house. Later he told me that the job couldn't have come at a better time. Several months before I called him, he had done extensive work at another house. The owner not only failed to pay Herb for his work but he sued him for damage he alleged Herb had done to his property — a complaint I find

Here comes the master excavator, Herb Brockert, driving one of his larger investments.

hard to believe given Herb's artistry with his huge machines. "I think I'd have gone crazy during that period," Herb said, "if I hadn't had your house to work on."

The second issue I had to discuss with Herb and Tom that April morning was the septic field. When we shot some elevations for the house, we also shot some grades for the septic system. What we discovered was that the developer who built the access road into our property and to our nearest neighbor installed it at a lower elevation than the plot plan indicated. My septic field as originally designed was therefore going to be higher than the access road.

There were two ways to deal with this unwelcome revelation. I could rip out a major portion of the common access road and raise it to the level indicated on the plot plan. But that would make the access road steeper, and in order not to have a big bump where the road came to my septic field, I'd have to regrade almost up to my neighbor's garage. It would be a major expense. The disruption would greatly annoy my neighbor.

If I didn't re-engineer the access road, the only alternative was to explore the possibility that the area designated for the septic field would tolerate a deeper field than the existing septic system plan called for. In 1985, when the excavation was done for the deep hole and water percolation tests, excavators dug down only seven and a half feet. "They probably hit a large boulder between seven and eight feet down," said Herb. "Or, if it was a small machine or a wet year, they may just have stopped once they got to the depth they thought they needed."

Since we hadn't yet determined the final elevation of the house, which in turn would influence the best available or even possible level of the septic field, we didn't have to do or decide anything about this issue immediately. I just tucked it into the back of my mind to think about when I had a restless night and lay awake worrying about unresolved issues.

Driving his excavating machine — a kind of giant backhoe — which he handles with the agility I bring to a power tool, Herb began to make an access road along the path of the driveway, pulling stumps as he came to them. He stripped off the topsoil from the driveway and dumped it in a pile near where the well would be dug; later it would be valuable topsoil cover for the lawn and gardens around the house. Once the drive was cleared, Herb removed all of the remaining stumps on the building site. Temporarily, the stumps were put aside in what became a huge pile.

When it is feasible, a builder digs a pit to bury stumps of felled trees right on the site. The builder of my neighbor's house to the west had done that, though he had dumped them down a bank at the back of the property and pushed a little fill over them rather than dig a proper pit. Maybe he, like me, didn't have a good place to dig a stump pit. On our plot, once one discounted the wetlands and buffer zones on both ends, and the construction site and septic field in the middle, there wasn't any stump burial place free of sub-surface granite. It isn't wise to bury them under a driveway because as the stumps deteriorate underground, sinkholes are apt to develop. The stumps would have to be hauled away.

Finding a legal stump dump site is getting more and more difficult because of environmental restrictions. But Herb found one about ten miles away. There were three tractor trailer loads of stumps! To make the operation more economical, each vehicle carrying away a load of stumps brought back a load of gravel for the base of the driveway. I think there's at least a foot to a foot and a half of gravel in the driveway base; the gravel replaced the topsoil that had been removed and brought the driveway back up close to finish grade.

When the driveway was operational, Herb turned his attention to the foundation. Again, he first stripped off the topsoil from the house site and added it to the pile from the driveway excavation. For the foundation work, he brought in a front loader to augment his excavator — total cost, $1,200 per day. The cost doesn't seem unfair given the value of the equipment involved, but it does make one anxious to have the work progress swiftly.

The minimum depth of a foundation varies from region to region. On Cape Cod, for example, with its sandy soil, the minimum depth is only eighteen inches below grade. But where Laura and I were building our house, we had to go down a minimum of four feet to get below the frost line. Where water tables are high, close to the surface — parts of Cape Cod, on Nantucket, where I was contractor for several projects, and in parts of Florida, for example — people often prefer to build slab foundations as close to the surface as possible because keeping full basements dry is simply too much of a hassle. But then the mechanical systems must be on the first floor, using valuable living space.

Since our foundation had to go down a minimum of four feet, we would have at least a crawl space under our house. But to me, there's something unsatisfying about crawl spaces or partial basements. They are difficult to ventilate properly year-round. If they aren't ventilated, humidity can build up during the summer, leading to mildew and, eventually, dry rot. Crawl spaces are difficult to work in. Plumbers and electricians hate to deal with them while installing or servicing mechanical systems. It doesn't cost that much more, usually, to go down another four feet to get a full basement. My philosophy is, if you're going to dig at all, you might as well dig all the way.

Events, however, were quickly going to challenge my philosophy and force me to make some compromises. About two feet under the surface of where the north end of the main house would sit, Herb ran into a solid granite ledge he couldn't budge with his excavator. It was just where he suspected he might hit subsurface rock. There was also some subsurface sedimentary rock he couldn't dislodge along the north wall of the ell and along the north end of the office wing.

How much should we try to blast out? Jock Gifford suggested that we be satisfied with a full basement under the ell and a partial basement under the main house. We wouldn't try to break up the solid granite ledge under the north end of the main house. The space under it would have to be one of my maligned partial basements. The sedimentary rock under the ell and garage, however, would be easier to break up. Herb Brockert undertook to hire a demolition crew to do the blasting; until they arrived, he continued to excavate where subsurface rock was not a barrier.

There were two sessions of blasting. For the first session holes were drilled by an air-powered tractor and drill that could be operated by remote control. The machine drilled thirty holes up to twelve feet deep through rock. Once the last stick of dynamite had been wired and placed, the blasters, their shirts and jeans stained with a mixture of sweat and mud, scattered. Most jogged behind their truck, instinctively ducking. Others crouched behind a boulder on the knoll.

"Fire in the hole!" yelled the foreman as he pushed a small lever. A rusted metal mat, woven like a rug, lurched a few feet into the air where it covered the blasting site, then fell back with a thud. The ground shook briefly, and a sharp crack echoed through the woods. Herb watched approvingly. The blasting wasn't his trade, but it fascinated him. His equipment had been part of the blasting operation because he used his excavator to place the heavy mat over the blasting area and to remove it after the explosion.

When the debris of the blast had been cleared away, we decided that a second, but lesser, blasting session would be necessary to remove more rock along the back wall of the ell. It wasn't a large enough job to require bringing back the air-drill machine. The blasting crew used jackhammers powered by a compressor to make the holes for the second blast.

When Herb had removed the debris of the second blast, he saw there was enough space to allow a full basement under the ell. But there wouldn't be a full basement under any of the main house. The north end would have a crawl space under it about two and a half feet high; the space under the south end would be about five feet high, plenty high enough for the installation and servicing of utility systems. The cement-floored space would be open to the ell basement, allowing for good air circulation. I didn't expect to have any venting problems.

Once Herb had buried all of the blasting debris as fill, I could see little evidence that we had done any blasting. In order to get as much crawl space under the main house as we did wedge out, I raised the elevation of the finished floor of the first story not once but twice — a foot each time — from 238.5 feet above median sea level to 240.5, which made the final elevation less than a foot higher than the highest natural elevation within the footprint of the house. Before Herb began the excavation, I was sure that setting the house as high as 240.5 feet would make the south end of the main house loom up out of the ground in an awkward way. After Herb had cleverly distributed the excavated subsoil and rock around the site, it looked very different. He was able to reduce significantly both the north-to-south slope and the east-to-west slope. The vicinity of the excavation was beginning to look more like a level lot.

Before he moved his heavy equipment off the property, Herb piled some of the reserved subsoil close to where the foundation walls would be poured, so that it would be relatively easy to backfill against the concrete foundation walls once their forms were removed. He also made a packed-down earth ramp from the driveway up to the edge of the foundation for the convenience of concrete trucks backing in to dump their loads.

With his excavator on a truck ready to move out, on June 11, 1992, Herb said, "Well, with a property like this, you never know. It wasn't so bad." Not bad, he was saying to me, but as he pulled away I was adding up the costs in my head. The site, neatened up, didn't look as though all that much had been done to it. Trees and brush had been cleared, there was a rough driveway, and there was a hole for the foundation. But we had put close to $20,000 into the preparation of the site, and we hadn't even poured a foundation yet.

A Well and a Permit

 WE HAD JUMPED the gun by clearing the site and digging a foundation before the town had issued a building permit. But we couldn't do any construction until I had the permit in hand. Two of the three elements I needed to qualify for it were in my possession: architectural and engineering plans for the house, including the timber-frame design of the ell; and the permit issued by the Conservation Commission related to our variance for encroaching on the buffer zone of a wetland. What remained was to get a well driven and certified. No well, no house.

Properly speaking, when a drilling rig rather than a shovel is used, a well is "driven" rather than dug. It didn't take nearly as long to drive a well as to dig a foundation. Less than a day. My schedule kept me elsewhere that day, June 19, 1992 — I never saw any of the operation. Dave Haynes, another subcontractor familiar to me from *This Old House* renovations, came down from southern New Hampshire with his rig. He had to have two separate permits from the town: one permit to drive the well, another permit to install the pump — each permit with its own fee, of course.

Since our town lacks a municipal water system, there is no option for homeowners other than driving a separate well for each property. For convenience, our well was going to be about twenty yards from the southeast corner of the garage. In case there was ever any trouble in the underground

Air forced down through the drilling mechanism causes water standing in the shaft to spout dramatically when it is evacuated. Dave Haynes beats a fast retreat from our version of Old Faithful.

pipe running from the well to the house, I didn't want to have very much excavation to do to replace the line.

My first question to Dave was, "Do you think you'll hit water where I staked out the well?" Many of us think it matters where someone drills for water, but according to Dave, we're all mistaken. No matter where he puts the drilling rig down, he's going to hit underground water. He won't guarantee how much water or how pure, but at some depth he'll hit it.

On average, a well driver will hit underground water in our town at about 200 to 230 feet. Dave hit it on our site at 187 feet. His drill bit makes a shaft about six inches in diameter. Where the drill passes through soil he drops casing into the shaft to keep the shaft from collapsing. But where the drill bites through rock formations, the rock itself is a perfectly good natural casing. Where Dave drove our well, all but about thirty feet went through granite or other rock, so he didn't need much casing.

Pat Loring from the Conservation Commission came by on a routine visit while Dave was driving the well. She was very impressed that having noticed he was working close to a silt barrier protecting a wetland and its buffer zone from a construction site, Dave had dug a pit around the drill pipe to catch the driving chippings, the almost microscopic pieces of drill bit that chip off as the drill drives through rock. "A lot of well drivers," said Loring in admiration, "don't know what we're talking about when we discuss the ways a drilling operation can contaminate a site."

The water supply Dave hit yielded a flow of eighteen gallons per minute, more than sufficient for our needs. We installed a pump that can draw sixteen gallons per minute, less than the available supply. The testing procedure for a new well requires the driver to pump it continuously for several hours to make sure that any contaminants from the driving are flushed out of the shaft. Then a sample is taken and delivered to an independent testing laboratory.

Since I didn't know how long the well inspector for the town would take to approve the well, I personally delivered the test results to the town hall and said, "You've got to sign off" (meaning, issue a certificate of potable water). The day I got written approval of the well I took it and the other documents I needed, and I went to get the building permit.

Bob Koning, the courtly white-haired building inspector for our town, looked through my documents and plans with interest, asking intelligent questions. He noted that most builders are using two-by-six studs in their walls, but I was showing only two-by-fours except in the north wall of the ell. I explained that using the lighter studs would enable me to improve the quality of insulation.

When Koning looked over the engineering specifications for the timber-frame ell, he whistled in admiration. "Way over code," he said. "Code" was a shorthand reference to the 850-page commonwealth building code. For a long time the Commonwealth of Massachusetts left building regulation mostly to local communities. Before 1975, 287 cities and towns had their own building codes, supplemented by eighty-seven commonwealth regulations. It was a nightmare for contractors doing work in several communities at the same time.

In 1975 Massachusetts issued a uniform commonwealth code replacing all of the local codes, but over time the code had grown to intimidating length and complexity, much of the growth caused by developments in construction technologies, materials, and safety regulations. The town building inspectors could shut down a project for any perceived violations of code, and they could require that any work that fell short of code requirements be redone before a certificate of occupancy was granted. "People come in," Koning said as he computed the cost of my permit, "and want to build here with plans drawn by some company in Georgia. They think they can build exactly the same house in New England that is suitable in the South, and when something goes wrong, they want me to rescue them."

The building permit cost $1,888, about what I expected. "When you're forming up [setting up the forms for the poured concrete foundation], give me a call," Koning said, dismissing me cheerfully from his office in the town hall. In my hand I held building permit number 4970 dated July 6, 1992, printed on burnt orange colored cardboard; it notified the world that the "petitioner [Laura and I] has permission to build a new dwelling."

How Firm a Foundation

It WAS A Thursday afternoon in late July of 1992. My foundation contractor, Phil McCarthy, and his crew and I were stewing more or less silently, waiting for the town building inspector. There were at least eight days of work ahead for this crew. Phil was itching to get going because the local economy was improving; he had other jobs lined up to move on to once he got our foundation up. I was itching to get a foundation up so I could bring framers in.

The crew had already set the plank forms for the two-foot-wide concrete pads called "footings" that lie between the ground and the foundation walls. They had backfilled the planks with soil and rocks to prevent concrete from escaping through any gaps in the forms while it set up. Where drainage pipes were to pass through the footings, they had inserted tin sleeves. They had unloaded from Phil's truck an extension chute that they could use to give more length to the chute on the concrete truck. With the extra-long chute they could make concrete flow to hard-to-reach forms rather than moving it there with a wheelbarrow and shovels. All was ready, but the building code did not permit us to pour any concrete until the building inspector acknowledged that the footings were being set on good bearing soil.

Shortly before three o'clock, Phil sent the crew home. They had run out of things to do. The tools of their trade — long-handled shovels, mallets, a dusty transit — rested against a pile of gray forms. Phil and I both knew that the concrete supplier made no deliveries after three o'clock. By five-thirty Phil and I finally had a firm commitment from Bob Koning as to when he would check the footings: on the following Monday! I bit my lip in frustration. I was quickly relearning the trials of coordinating the schedules of different parties — in this instance, the foundation contractor, the concrete supplier, and the inspector — who often seem to be pulling in opposite directions.

At our site, Bob Koning had to approve both the soil under the footings and the system we had devised to secure the foundation to the top of the granite ledge in the crawl space. Phil and I had studied the ledge. It wasn't level; it had some pitch to it. This section would be Phil's toughest pour. A cubic yard of concrete can weigh four thousand pounds and, as Phil reminded me, "it likes to slide downhill wherever it can." There was the problem of holding it in the right place while it set and the problem of holding it in place after it had set.

To prevent the foundation on top of the ledge from ever slipping after it had set, I drilled a row of holes into the ledge along the line of the foundation. Then we inserted steel pins into the holes in such a way that they extruded above the holes. When the foundation concrete was poured around the pins, the interlocking of the pins with both granite and concrete guaranteed the foundation would not shift.

I spent one long day drilling holes into the granite myself. "How many holes?" Phil asked, staring at the steel pins sticking out of the ledge as we waited to hear from the inspector. "I don't know," I admitted. "I didn't count them. I just kept drilling until the bit got too dull to bite through the granite anymore."

Phil had quickly won my confidence. I didn't have any leftover contacts with foundation contractors from my days as a general contractor. But I knew someone who crossed paths with them frequently — Herb Brockert. As excavator, Herb was on the scene just before the foundation contractor to dig the hole, and again just after, to backfill the poured foundation. Herb recommended Phil to me. I computed the number of cubic feet of concrete required for our foundation as part of a calculation of what the foundation should cost. Phil's bid was well within my range.

When I met Phil he was going to school part-time to finish his engineering degree. Handling foundation concrete is heavy work best suited to young men. Phil was still fairly young and very strong, but he had wisely foreseen that there was a better future for him in middle age than muscling concrete.

Poured concrete is the preferred foundation in New England. The heavy clay and the moisture in the soil create pressure against a foundation that the strength of concrete can resist. The mixture of sand, cement, and three-quarter-inch stone (further strengthened with steel-reinforcing bars called "rebar") that Phil was ordering is calculated to withstand up to three thousand pounds of pressure per square foot. Given proper drainage around a poured concrete foundation, there will never be a problem with it for the lifetime of the house.

Bob Koning came as promised on Monday, early enough that we poured footing in the afternoon. While the concrete was still wet, the crew carved grooves called "keyways" on the top surfaces. When the concrete is poured for the wall sitting atop the footing, the concrete flows into the keyways and the wall then cannot slide on the footing.

Forming up the foundation walls was next. Foundation contractors generally bring their forms with them. (Phil's forms were made of plywood with two-by-four frames; the largest were eight feet high and about thirty inches wide.) The forms clamp together as needed to make the mold for one side of a foundation wall. A series of tie wires between the two lines of panels that contain a poured concrete wall prevent the panels from bowing out under the weight of the heavy liquid concrete before it sets. The crew break off these wires when they remove the forms, leaving most of the wire embedded in the concrete.

I snapped the chalk lines myself to indicate to the foundation crew precisely where the walls should sit on top of the footings. Then they installed the forms for the long north wall of the main house, ell, and office wing. They put two steel bars at the bottom inside each form, and two at the top, running horizontally the length of the wall to reinforce it. Concrete is very strong in compression — that is, when it is holding something up. But it is weaker than wood of comparable thickness under tension — when it is being pulled apart or bent. The steel rods give the concrete added strength to deal with tension.

Plumb Bob
A simple tool that saves a thousand mistakes.

Phil McCarthy sprays a mixture of oil and water onto the inside of foundation forms just before concrete is poured to inhibit the concrete from adhering to the forms. How do we know this is a contemporary picture? Because Phil is wearing a beeper.

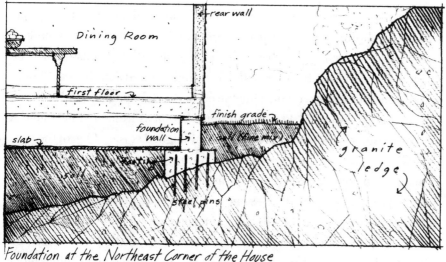

Foundation at the Northeast Corner of the House
The foundation needed to be anchored into the granite ledge.

The crew reinforced the outer surfaces of the forms with wood braces called "whalers." They also "kicked the corners," adding scrap-wood planks and extra forms around any corners, where poured concrete will exert the most pressure against the forms. Phil's plan was to pour the foundation in a few stages, leaving the south side until last. The driveway is on the south side of the excavation. The mixer trucks could back all the way inside the foundation to pour the other walls at close range so long as the south wall was open. Then, when the north wall had set enough for the forms to be removed, Herb Brockert could bring his big excavator machine back inside the foundation and backfill the outside of the wall from the inside. The proximity of the wall to the knoll meant he could never get his machine along the wall from the outside.

Before the pour, the standing forms, braced as they are by metal clamps and wood two-by-fours, look almost permanent. Actually, they are fairly weak as a structure, especially on a long run like the north wall. You can lean your weight against one point of the wall of forms and see it wobble. What makes the structure strong is filling its core with concrete, but pity the poor contractor who lets the wall of concrete set after the forms have shifted away from where they're supposed to be.

Using a "plumb bob," a weighted line that measures a vertical plane, Phil discovered one corner of the forms to be a quarter of an inch off vertically straight, or plumb. "Gotta pull it a quarter," Phil shouted to a crew member. Grabbing an oversize crowbar, which he called the "monsterbar," he jammed the bar under the base of the form and grunted and shoved. "There you go," said his helper, rechecking the corner and finding it plumb. "It's all mathematics," shrugged Phil. Continuing his measurements, he found a more worrisome problem than the single misaligned corner. One of the fastest and best ways to judge whether a rectangle or square is, as we say, "square" is to compare the diagonal distances of the area's four corners. If the two diagonals are exactly the same length you have a rectangle or square, but if they differ you have a parallelogram and a problem! Phil found a two-inch difference between the diagonals. Worried that the forms might be shifting as the crew went about its business — "I've got too many things moving," he said — Phil continued to add braces. Finally the forms were plumb and square. Next, using a transit, Phil set elevation marks for the final height of the poured concrete at several places along the inside of the forms. A chalk line would be snapped between these points and then nails driven partially into the forms along the chalk line to act as a guide to ensure that the concrete was poured up to the correct height. "Money!" — meaning "It's on the money" or accurate — he shouted, peering through the eyepiece of his transit after each sighting had located the final height of the foundation.

Shortly before the pour, the crew sprayed the inside surfaces of the forms with a mixture of oil and water, a "release agent," to make the surfaces slippery to prevent the concrete from bonding to the forms as it set. The first truck of concrete backed into place. The driver got out and activated a lever to accelerate the spin of the huge drum of concrete. Before he released any concrete into the chutes, Phil checked the consistency of the mix. A sample of the mix is poured into a cone and the cone inverted on a flat surface so that the conical sample is standing on its own; the extent it collapses or "slumps" from its original height is measured. On a typical commercial pour, where the walls and slabs tend to be more massive than is required for a typical house, and where the immediate strength of the concrete may be more important, a fairly dry three- or four-inch slump might be ordered.

"Start it at five," Phil told the driver over the din of the churning drum.

Occasionally, the truck could get close enough to the pour to use a single chute. More often, extension chutes weighing forty to fifty pounds each had to be added. The cement has to be worked in the form as it is poured to prevent honeycomb-rough texture on the surface, and to ensure that it gets around the rebar and into all corners, leaving no voids.

"Then maybe we'll run it to six." To move the liquid concrete the length of the long north wall, Phil knew he would need a pretty soupy mix. He and two of his crew were just behind the truck, balancing themselves on top of the forms, eight feet above the floor of the excavation. As they unhooked a portable twelve-foot steel "halftube," a metal extension on the truck's permanently attached main pouring chute, the main chute, freed of the weight of the halftube, jerked upward. It hooked the shirt of one of the crew, lifting and pulling him at once. I saw his eyes widen and his face freeze with tension. Then the chute stopped, and, for a moment, the crewman lost his footing. He began to fall backward but grabbed the main chute with both hands. Only his handhold kept him from falling backward into the excavation.

Phil, holding the halftube, lost his balance, too. The second crewman, muscles straining, kept the freed halftube from falling, taking Phil with it. In a second the incident was over. Phil recovered his balance. The first crewman got his feet back under him and his shirt unhooked from the chute. No one fell. No one said anything. They continued with the pour. Only the first crewman's occasional rubbing of his chest where the chute had struck him as it hooked his shirt suggested that anything unusual had happened.

When the concrete began flowing, Phil and his crew moved nimbly along the tops of the narrow forms. They kept checking the forms carefully to make sure nothing shifted even slightly from where it was supposed to be. The goal is for the finished wall not to be more than an eighth of an inch off line. Pouring, moving the forms, pouring another section continued for several more days with stupendous exertion of physical energy but no accidents.

Concrete, though dense, is somewhat porous. I've read that as many as 40 percent of new houses have leaky basements. Proper excavation and backfilling should eliminate much of the potential for water penetrating the foundation. I took steps to keep our basement and crawl space dry. After the forms had been removed from the concrete foundation walls, I damp-proofed the outside of the walls with an asphalt coating. Many people think this coating waterproofs the foundation wall; but it only stops dampness from working through the wall. If I had wanted a totally waterproofed foundation I would have had to hire another subcontractor to apply a fully waterproof membrane to the outside of the foundation concrete. Because I would have good foundation drainage, I didn't think I needed complete waterproofing.

To keep water out of the basement we laid perimeter drain pipe along the footings of the north side of the house — the side that would be subject to surface water running or trickling down to the house from higher surrounding elevations, notably, from the knoll that rose ten to fifteen feet on the north side. The four-inch perforated PVC (polyvinyl chloride, a water-insoluble resin that is highly resistant to chemicals and corrosion) pipe was laid in about a foot of washed stone. These drains would come into play, I judged, only in the event of a sustained monsoonlike rain. The pipe was pitched to carry any such water around the house and release it to daylight.

I also installed perforated PVC pipe along the inside perimeter of the footings in the basement and crawl space. When we filled the foundation floor with crushed stone up to the top of the footings, this pipe was surrounded by the stone. It served the double purpose of draining off any water collecting under the basement floor after a heavy storm and of venting any radon gas in the vicinity. Radon is the invisible but toxic product of minute traces of uranium, which may be found in ledge formations like the one we were building on. We wouldn't be able to test for radon until the basement was enclosed. If there proved to be any traces of radon, I could remove and disperse it by connecting a fan to the stub we brought from the pipe up into the utility room.

Before Herb Brockert returned to backfill around the foundation, I first had to install insulation on the outside of the foundation walls. Foundation insulation isn't code-mandated everywhere, but in any location with cold winters it's a pretty good idea. Many existing homes lack it. Basement furnaces were once relatively inefficient; homeowners counted on their throwing enough heat around them to keep basements warm. Furnace technology now emphasizes efficiency. One of the renovations I did in our old garrison colonial was to install an 87 percent efficient European furnace recommended by Richard Trethewey, the heating consultant on *This Old House*. Little heat is then thrown off in the basement. Since the foundation of that house was uninsulated, most of the heat that was thrown off in the basement radiated through the walls. The basement always felt chilly.

On the outside of our foundation I installed two-inch-thick extruded styrene foam panels with an R-10 insulating value. Some builders advocate insulating the foundation walls right down to the footings. I applied the foam panels only to the top four feet of the walls, which provided insulation from grade level down to below the frost line in our area. Below that point, the ground has essentially constant temperature.

Herb Brockert then backfilled the foundation with a loose, gravelly mixture of fill, so that any surface water would not get trapped around the foundation but would settle to the perforated pipes at the footing level and drain to the release point back of the garage.

It was not until the entire foundation had been poured and backfilled that I could form a first impression of how the house was going to fit into its site. Would it loom too high? Had I miscalculated — would it sit too low? As I drove around the bend of the driveway each day, I could picture the top edge in silhouette. It looked good to me.

Amityville, Massachusetts?

 **DURING THE CLEARING** of the site and the installation of the foundation, Laura was an occasional visitor to the site. I oversaw the work, dealt with the subcontractors, ordered materials, wrote the checks, and, when my schedule permitted, did some of the work.

One afternoon in late July of 1992, while I was installing insulation on the foundation walls, Laura drove in, accompanied by her friend Deanna. Deanna, Laura said, was an expert in the realm of spirits. She had offered her services when Laura confessed to being bothered by the old, open granite-and-fieldstone–lined foundation in what was going to be our front yard.

Laura had never changed her opinion that the open foundation was spooky and should be filled in — the sooner, the better. Tom Wirth, our landscape architect, was as enthusiastic about the foundation as Laura was apprehensive. He saw it as a beautiful sunken garden site. On this issue, I agreed with Tom. I'm not about to say that nothing would ever spook me, but the old foundation really didn't bother me.

To Laura, this hole was the Amityville horror revisited, a haunted place. She remembered trees shaking and lightning flashing as Margot Kidder dug in the backyard of the Long Island house, trying to identify what supernatural force was trying to force her family to abandon their house; and she imagined herself dealing with unfriendly local spirits while I languished hundreds or thousands of miles away giving talks or signing autographs at a home show.

Deanna steadied herself with a stick as she negotiated the ruts and rocks between the rough driveway and the old foundation. When she and Laura climbed down into the four- to five-foot-deep rectangular hole it was the first time that Laura had actually been inside the foundation. "It's certainly not spooky," Deanna said. "I can sense if it's spooky."

"I don't believe anyone ever lived here," I told Deanna before she descended into the stone-lined excavation. "This foundation was probably the support for a shed when they quarried granite on this property." "A hiding place?" mused Deanna, looking around. "Guns . . . a hiding place . . . for soldiers, or people. It hid something." Well, I had read in a local history that one of the colonial Minutemen once lived nearby — perhaps a quarter of a mile from our site — so I couldn't definitely rule out what she was sensing.

"Dead bodies?" asked Laura. "No dead bodies," Deanna assured her. They climbed back up out of the foundation. I mentioned to Laura that after we moved in I would probably dig around to see if I could find any colonial or Indian artifacts. "If you find any bones," Laura said, "don't tell me about it."

Then Deanna surprised me with the biggest bonus of her visitation. Looking at the excavation for the house and the lines of the foundation walls, she said, "What a big house!" For a couple of months I'd been waiting for Laura to come to the same conclusion. We were building a house with about 4,600 square feet of interior space, more than 1,600 square feet larger than our old house. Even before we began clearing the property I had walked Laura through our old house with a tape measure and shown her how much larger the new house would be, room by room.

"You can see that?" Laura asked Deanna. "Doesn't it look dinky? How will we fit all of the rooms in? I can't picture it." But Deanna had already turned her attention to the tract of dense forest beyond the stone wall along the south border of our plot. "I feel a lot of Indians here," she said, "and that goes way back."

Returning to the car, Deanna stopped once more. She looked toward the west slope, where the lot drops off down to the wetland and the trickling stream. "Keep walking straight out that way," she said ominously, pointing westward. "You might find something." That night, over dinner, Laura asked me what I had found when I investigated where Deanna had pointed. I had to confess that I found nothing because I hadn't looked. Feeling the pressure of getting more insulation on the foundation walls, I had hurried back to my work as soon as Laura and Deanna drove away.

"One day you'll leave on one of your television or personal appearance trips," Laura half warned me, "and I'll post a sign. 'Topsoil wanted.' " She didn't have to explain where the topsoil would be dumped.

Raising an Eyebrow

PHIL McCARTHY and his crew waited until the concrete in the foundation footings and walls had cured for a couple of weeks; then, in mid-August of 1992 — after Herb Brockert had put down a crushed stone base from the ground up to the top of the ten-inch-thick footings — they poured concrete for the basement under the ell and the crawl space under the main house. Phil put some "control joints" in the concrete slabs where he thought the slabs might be subject to stress as they dried, for example, where a floor turned a corner.

A control joint is supposed to draw any inevitable cracking to itself and keep the crack in a straight line, thus eliminating random hairline or wider cracks. The old method was to hand-groove these joints into the surface of the fresh wet concrete. Phil used plastic T-shaped strips for the joints. He buried them with the top of the T flush with the surface of the concrete. Then he pulled off the tops, leaving the tail strips buried in the concrete. Around the perimeters of the slabs he inserted expansion joints made of rigid foam to absorb any seasonal expansion or contraction action between the slabs and the foundation. If the joints ever had to be sealed — after finding evidence of radon around the floor or foundation, for example — the upper portion of the foam would pull out easily, leaving space to apply a tight caulk seal. Phil also sprayed a polyurethane coating on the concrete floors to slow the curing process in the interest of getting a stronger floor, and also to seal and protect the surface of the concrete.

The foundation now looked complete, but it wasn't. The Massachusetts building code requires that at least six inches of the top of the foundation be above grade to keep ground moisture and pests in the soil from having direct contact with any wood framing or sheathing. I know it's common just to ignore that strip of exposed concrete, or maybe to camouflage it with foundation plantings, but it detracts from the appearance of a house. If you look at old houses — and the facade of ours was intended to look no more recent than eighteenth century — you will see attractive strips of exposed foundation made of brick or stone, traditional materials.

Before Phil even poured the foundation, I had decided to add a masonry cap to bring the foundation above the finished grade. The cap would be concrete block narrower than the foundation — leaving a ledge on the outside on which to rest a veneer. None of the poured concrete wall would be visible outside. Jock Gifford sketched a veneer made of brick. Russ Morash, my television producer, cautioned against using brick. "You know," he told me, "if you use brick, you will have all those mortar joints, and the more joints, the more opportunity for cracks to open, through which moisture can penetrate. It's like a bathroom shower. The smaller the tiles, the more likely they are to crack."

What appealed to me, building on a site where granite had once been quarried, was to dress the exposed strip of foundation with a veneer of thin slabs of granite. Full of enthusiasm, I went to a nearby quarry and priced pieces of granite with the dimensions of an inch thick, a foot wide, and five to six feet long. The quotation destroyed my fantasy. The milling necessary to make pieces that thin is expensively time consuming. The wafer-thin pieces might break easily; the amount of breakage even in careful installation could make a costly item even dearer. Also, the pieces would be very smooth. I thought a little more texture to the stone would look better.

At this juncture, I telephoned Roger Hopkins. A wizard with all kinds of stonework, Roger has created and installed several kinds of stonework projects before the cameras of *This Old House.* "Why not use curb stone?" Roger suggested. "It's readily available and not too expensive, about a fourth

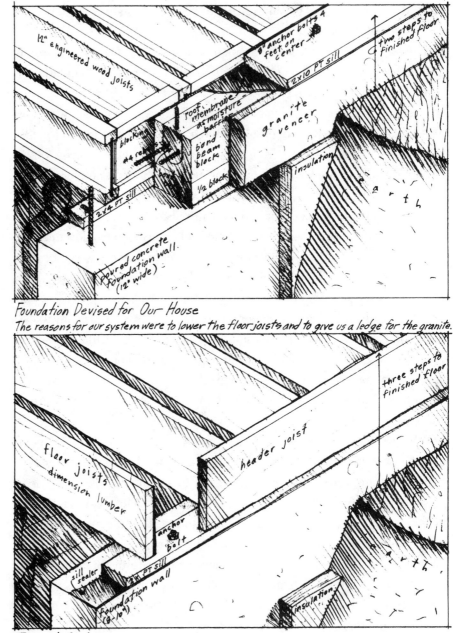

12" engineered wood joists

8" anchor bolts 4 feet on center

2×10 PT sill

two steps to finished floor

roof membrane as moisture barrier

blocking

#4 rebars

bend beam block

granite veneer

½ block

insulation

2×4 PT sill

earth

poured concrete foundation wall (12" wide)

Foundation Devised for Our House
The reasons for our system were to lower the floor joists and to give us a ledge for the granite.

three steps to finished floor

floor joists dimension lumber

header joist

anchor bolt

sill sealer

2×8 PT sill

foundation wall (8-10")

insulation

earth

Typical System

of the cost of the milled granite you priced. The pieces are about three inches thick, but maybe we could split them in half, or just select some of the thinnest pieces. Curb stone has the slightly rougher finish you prefer." Jock Gifford endorsed the idea and suggested that I rely on Roger to devise a method of attaching the granite to the concrete block and foundation.

To complicate matters even more, I wanted the house to sit low enough to the finished grade around it that no more than two steps would bring a person from the front walk into the front hallway. This would not be possible if the first-floor joists sat as usual above the foundation or even on top of the cap. Our design called for another ledge on the inside of the foundation to carry the floor joists.

Since we knew how the top of the foundation was going to be finished off before we poured it, Phil and his crew installed a number five steel "rebar" (a bar ribbed to give it greater bonding strength when used to reinforce concrete) every four feet along the center of the top of the wall; one end of each rebar was anchored in the concrete, the other end stuck up several inches above the concrete. Phil also set anchor bolts every four feet, about an inch in from the inside edge of the top of the foundation. I snapped chalk lines on top of the foundation to mark where a sill of pressure-treated two-by-fours should be positioned on the inner edge of the wall top. I drilled holes in the two-by-fours where the anchor bolts were located and secured the sills to the foundation with the bolts. The ends of the first floor joists were going to sit on the sills in due course.

It meant a lot to me to lay the first pieces of wood for our new house. Other carpenters were going to cut and install truckloads of wood before we moved into the house, but I had the satisfaction of laying the first pieces, and eventually, in finish details, I intended to lay the final wood myself.

Lenny Belliveau, our masterful, soft-spoken mason of French Canadian ancestry and birth, took over at this point. He laid a course of cement half-blocks along the middle of the foot-wide foundation wall, setting the blocks in a bed of mortar so that they were right up against the wood sill I had installed. Then he laid a course of "bond-beam blocks" in a bed of mortar on top of the half-blocks. Where the rebar stuck up, he drilled holes in the blocks so that they fitted down over the rebar. The bond-beam blocks are U-shaped. Where the tops of the rebar emerged into the open mouth of the U, two more steel rebar, number fours, were laid horizontally, one on each side of the vertical rebar. Lenny wired the three rebar together. He filled the trough of the U with concrete and, while it was wet, embedded half-inch anchor bolts in it; eventually these bolts would anchor another sill above them.

Now we had a foot-high cement block wall sitting on top of the poured concrete foundation wall, but Lenny's little wall was narrower than the foundation under it, leaving a ledge on each side. I had already attached the sill of two-by-fours on the inside ledge as a bearing point for the floor joists. The outside ledge was available to Roger Hopkins as a place on which to rest the pieces of granite veneer. After Roger finished installing the veneer, a two-by-ten sill of pressure-treated wood, held in place by the anchor bolts that Lenny installed, would extend out over the granite veneer.

Roger brought forty-five pieces of granite to the site, each piece four and a half feet long, eleven and a half inches wide, and about three inches thick. These slabs had to be "dressed" — trimmed and shaped — so that they would be flush with each other along their exposed surface. They were intended to look as though they were the exposed sides of massive blocks of granite holding the house up, so Roger couldn't let any part of the mounting device show.

With his wild hair pulled back with a bandanna, bare to the waist in the heat of August 1992, Roger looked like a pirate as he wrestled the sixty-pound pieces of granite. The process was methodical and slow. Roger would measure each piece, then trim it, if need be, by scoring it with his grinder, knocking unwanted segments off with his mallet, and finally cleaning it up with his chisel.

On the bond beam I'd written with a lumber crayon "face of granite $2\frac{3}{4}$ inches from block." Working with his son, Jake, Roger drilled a shallow half-inch-wide channel called a "kerf" across the tops of both the granite slab and the bond beam next to it. He laid a piece of heavy wire in the kerf, anchored one end of it in a hole he drilled into the bond beam, then anchored the other end in a hole drilled into the top of the granite. The wire prevented the piece of granite from tipping away from the bond beam. To attach the granite to the bond beam even more securely, Roger poured a soupy mortar mixture into any small gaps between the bond beam and the granite.

*Roger Hopkins trims
the end of a piece of
granite veneer with a
diamond-bladed saw
while his son, Jake,
hand-sprays the blade
with water to lubricate
and cool the blade.
Roger further trims
the end with a chisel
to refine the edge.*

With a small saw, Roger Hopkins cuts the groove or "kerf," which will house the wire holding the granite veneer against the foundation. Then he taps the wire into place with a mason's hammer.

Roger didn't have much of a ledge to rest the granite on. He would have preferred a ledge three and a half inches wide, but the ledge he had to work with wasn't any wider than two inches in some places. To balance some pieces, Roger sent Jake searching around the site to scout small sticks and rock chips — makeshift shims to give the slab enough lift in places to sit evenly on its concrete perch. "These are uncharted waters," he muttered.

Where the sill was narrowest, some of the granite rested on the sill but the outside of its resting edge was sitting atop the rigid insulation I had nailed to the foundation wall. "I don't like them resting there," Roger said as he troweled a bed of mortar he described as "strong with Portland" (heavy on cement, light on sand) for a piece of granite to rest on. After two hours of continuous work Roger had only a couple of the forty-five slabs in place. "The problem is," he complained, "this is an afterthought. But it's a builder's house, so what do you expect? I don't know what you've got against a good condo."

It may have been the heat that made Roger irritable about a plan that was anything but an afterthought. But I knew another reason why he was a little grouchy. He was working under time pressure of his own making, not mine. In two days he was scheduled to fly to Egypt to film a television program. He had been in Egypt once before, making a show about the methods used to build the great pyramids, and constructing a small replica himself. Now he was about to take on another stonework miracle of the ancient world.

It occurred to me that the engineers and foremen of the pyramids must have encountered a few problems that would have put our little granite veneer project to shame, but I wasn't sure that such an observation would lighten Roger's mood. I contented myself with reckoning that Lenny Belliveau was capable of picking up and finishing the granite veneer after Roger left.

As soon as I saw the first pieces of granite in place at the top of the foundation, I knew the veneer was perfect for the house. When Herb Brockert did final grading with fill and topsoil around the foundation, the topsoil would overlap the bottom of the granite veneer so that it would appear to rise out of the ground. The amount of overlap could vary as the terrain dictated. In places, almost all of the eleven and a half inches of granite would show, in other places, less, just so long as we conformed to the building code by having at least six inches of it exposed.

The day after Roger started attaching the granite to the foundation, Jock Gifford showed up at the site. Jock hadn't been there for weeks because I didn't need him to monitor construction. He thought the time and effort we had spent siting the house and tinkering with its elevation had been well spent. "The ell will get plenty of south light," he said approvingly. "And the terrace outside the ell will sit just high enough above the driveway going past it to the garage that if any cars are parked on the driveway you'll still be able to sit on the terrace and look straight over them to the beautiful tract of forest to the south."

The major purpose of Jock's visit was to discuss windows again. We had been talking about them off and on for months. Framing couldn't proceed very far unless we knew the exact size and placement of all the doors and windows. I was budgeting between $20,000 and $30,000 for all of the windows — a major expenditure — and I wanted them to be as maintenance-free as possible. Thus as many windows as feasible would be "clad" windows, which reveal wood frames from inside the house but have weather-resistant aluminum frames on their exteriors. The natural look of the wood on the inside would complement Laura's desire for a country colonial interior of exposed beams and wood floors. Outside, the sash would appear to be painted, but the aluminum would never have to have a brush applied to it.

Jock's elevation drawings for the main house showed double-hung units everywhere except in the all-seasons room. Each half of the double unit was divided into twelve panels of glass called "lights" — such windows are referred to as twelve-over-twelves. The wood strips dividing the lights in colonial windows are called "muntins."

Jock and I had been thinking and researching our way through a dilemma where colonial design meets modern technology. Classic divided-light sash have only a single layer of glass. But as concern for fuel conservation has grown in this century, single-glazed windows have given way to double- or even triple-glazed panes. Manufacturers can increase the insulating capacity of glass in any of several different ways. One is to spray coatings on the glass. Another is to suspend a sheet of reflective film between two layers of glass. More recently, manufacturers have begun using gases such as argon or krypton between the layers of double- or triple-glazed windows.

It is still possible to buy single-glazed divided-light windows, but they are so inefficient from an energy standpoint that I'd be ridiculed by all my construction buddies if I used them. It is also possible to get a custom-built window with double-glazing and true muntins between each light. Jock showed me an example. It didn't look right. The increased thickness of the pane to accommodate the double-glazing required a muntin that was an inch and a quarter wide — twice as wide as the muntin in a classic colonial window. The elegance of the classic window had been sacrificed for an efficient but awkwardly heavy look. The double-glazed, divided-light windows, besides looking odd, were so expensive that I would have shot my entire window budget acquiring only the windows for the main house.

In my opinion, the life of a good window ought to be at least forty to fifty years. Many of the high-tech products I see these days come with relatively

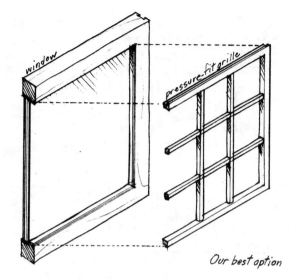

window

pressure-fit grille

Our best option

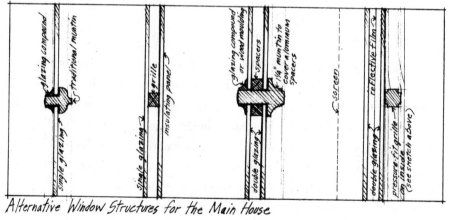

glazing compound

traditional / muntin

single glazing

grille

insulating panel

single glazing

glazing compound or wood moulding

spacers

1/4" muntin to cover aluminum spacers

double glazing

screen

reflective film

double glazing

pressure-fit grille on inside (see sketch above)

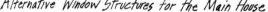

Alternative Window Structures for the Main House

Casement Window in Our Family Room
Both halves open inward; one half also tilts open.

short guarantees — ten years or less — and either fall apart shortly after the warranty runs out or are discontinued and become difficult to replace as needed. So the technology can be infuriating as well as progressive. On most clad windows, for example, you can't replace a single broken light. Once they're assembled, that's it. The only way to repair one is to replace the whole sash. At one hundred dollars and up per sash, you'd better be careful not to hit too many stray golf balls or backyard home runs.

Many window makers advocate creating the effect of divided-light windows by filling the sash with a single panel of insulated glass and then applying a removable grille to suggest the look of muntins. The trouble is that the grilles often don't do much more than suggest; they look tacked on and artificial. For weeks I worried about living with grilles. One manufacturer offered muntin grilles in a "designer" model. The grilles were glued onto both sides of the insulated glass so that only by examining the window closely would anyone ever notice that the glass wasn't bounded by each muntin but instead was a single panel of insulated glass that ran between two sets of fake muntins. The appearance was more convincing than other grilles, but the series came ready-made only in a six-over-six configuration, and I definitely wanted the smaller lights of twelve-over-twelves.

Also, this manufacturer didn't offer the glass technology that I preferred — a reflective film suspended between two panels of glass. If I were to order custom-made twelve-over-twelves from this manufacturer, I would be adding a fancy overcharge to what was already a very expensive window; this solution was not within my budget constraints.

The sales rep for yet another manufacturer offered an alternative I hadn't seen before. Instead of providing a grille fastened to the sash with metal or plastic clips, a common solution that usually looks about as convincing as a typical hairpiece, his firm was marketing a pressure-fit grille, which, from the inside, could pass all but the closest inspection as a genuine muntin. Interesting. But what about the look from the outside? The pressure-fit muntin grilles were for the inside only; the manufacturer didn't provide any grilles for the outside of the window equivalent to the glued-on grille muntins for the windows I had rejected as too expensive.

Since our property includes wetlands on both the east and west ends, I knew that all of the windows that opened would need full screens to protect us from mosquitoes in summer. Full screens, I thought, would blur anyone's ability to distinguish from outside whether the windows were true divided-lights or grilles over a single panel of insulated glass, and whether there were grilles on both sides of the glass or only on the inside. The moral of this story is that the homeowner has to keep searching for the best option. The last window rep had offered me a product I hadn't been aware of. His window combined the highest available insulating value with the best price. I was learning how to live with grilles.

Once we had ordered ready-made windows for the main house, we turned our attention to the custom windows for the "walls of glass" in the all-seasons room and along the south wall of the ell. At first I pondered making these windows double-hungs as well, so that we could open them for ventilation. But I soon concluded that double-hungs would be impractical. The balance systems for double-hungs are pushed to their limits by weights of around seventy-five pounds. The size of the windows we wanted to install were near or beyond those limits.

Double-hung windows are customarily made of annealed or "float" glass. For windows coming down within eighteen inches of the floor as we intended for the glass in the window walls, the building code requires tempered glass that is heat-treated so it won't break into dangerous shards if someone falls into it.

Our eventual decision for the south wall of the ell was to put glassed French doors in the middle. The doors would be flanked on each side by three large fixed panels of glass. We discussed the possibility of using grilles on these large windows to make them look more like the twelve-over-twelve divided-light windows in the main house, but I thought it was preferable to leave these large windows plain — all the better to see the beautiful woods outside. The major reason for the wall of glass, after all, was to open the view to those inside the house.

The west wall of the all-seasons room meets the east end of the ell at a right angle. On the all-seasons west wall Jock Gifford fitted in two panels of glass with a single glassed door between them. The adjoining south wall of the all-seasons room is a little wider than the west wall, so Jock put French doors in the middle flanked by panels of fixed glass. The little rectangles of transom lights that run like a line of trim above the window wall of the ell and the two window walls of the all-seasons room make the entire span of three walls seem to be a single unit.

Jock found me a window and door manufacturer in Connecticut to custom-build these vertical picture windows of tempered glass framed in mahogany. He drove me down to visit the factory. I was impressed with the quality of work I saw and with the quality of hardware the company installed on its windows and doors. Although this company had not previously incorporated the specific insulating technology I wanted into any of its windows, the owners were happy to accept an order from me that was conditional upon their adopting it for our windows.

I ordered not only the windows for the ell but also the glassed mahogany doors for the ell and all-seasons room, and a couple of other exterior doors. Wood French doors can be drafty if the wood starts to twist as the doors endure temperature changes, but pattern-grade mahogany is very stable, and I believe the doors will not warp. (Pattern-grade is the very best quality mahogany, straight-grained, clear or free of knots — named for the wood used in the machine industry to make models that when cast in sand become the prototype for cast-metal objects.)

I was a little concerned that we might be mixing too many woods in the ell — the mahogany of the frames of the French doors and window units, the spruce of the trusses, the cedar of the posts and plate beams, and whatever wood we might choose for the kitchen cabinetry and the built-in shelves, cupboards, and window seat in the family room. Jock assured me the woods were all compatible with each other.

On the north wall of the ell, opposite the window wall, there were to be only two windows — a good-sized window over the kitchen sink, and an even larger window in the family room framed by built-in cabinetry and with a window seat below it. The whole style of the ell suggested a different treatment for these two windows from that of the divided-light windows in the main house. We decided on a simple casement window for the kitchen that would crank open outward. The family room would get a casement window with two halves that could crank open inward, but, for limited ventilation, one side could also tilt in about four inches from the top.

From the ell and the all-seasons room, a person can see only one window in the office wing. But what a window! — certainly my favorite in the whole house. This window is situated where the east wall of my office intersects the roof. It is mostly in the wall, but the top of the window rises above the eave line.

One way of dealing with a window that is partly in the wall and partly in the roof is to build a conventional dormer or half-dormer to extend the roof out over the window. What I wanted, though, was an eyebrow-like window. An eyebrow window is named for the way its curved top resembles the curve of an eyebrow. You often see eyebrow windows tucked into the attics of Victorian houses. Set into steeply pitched roofs, they usually are quite small and look uncannily like architectural eyes gazing out at the world.

Jock designed my office "eyebrow" window to have three sections — two casement windows flanking a section of fixed glass, all with curved tops. The

Eyebrow windows were introduced in shingle-style house roofs in the late 1800s to admit a little light into attics where dormers were not desired. I thought the curving top line would look graceful on the window of my office's east wall.

roof curves around the top of the three-section window the same way the roof curves around the top of a true eyebrow window, necessitating careful shingling around the curvature. This window interested me so much that I undertook all of the work on it myself except the actual manufacture of the window unit. I framed the inside curvature for the plasterer, I framed the outside, I did the shingling around the window, and I cut and installed the exterior trim.

Since the windows and doors now had specifications, there was nothing standing in the way of framing except completing the installation of the granite foundation veneer. Lenny Belliveau, always as mild-mannered as Roger Hopkins is sometimes cranky, came over as soon as Roger left for Egypt. He finished the veneer in early September 1992.

The Deck

DURING THE LATE summer of 1992 I went shopping for a car phone, thinking it might be indispensable for communication until we got a telephone installed at the building site. I mentioned to the salesman that I was building a house. He recommended a local carpenter for whom he had once worked during college vacations.

I didn't have a ready supply of candidates to frame the house because I had always done the framing myself for jobs that I managed as general contractor. But I didn't see how, with my workload, I could keep a regular enough presence at the site to oversee the framing myself. There was no reasonable alternative to hiring an unknown framing contractor who would have his own crew. But you can guess how I would feel if the job were not done according to my standards.

In the hierarchy of carpentry as a skilled trade, the ability to frame doesn't rank at the top. An aspiring carpenter, in my view, should begin as a laborer, progress to framing, move up further to exterior trim and siding, and finally graduate to finish carpentry. Fine millwork and cabinetry are the ultimate goals on the carpentry scale.

When my father taught me carpentry, our jobs began with the foundation and continued through the framing, roofing, and exterior trim and siding, culminating in the interior finish carpentry. There were no carpentry specialties. Framing carpenters, as specialists, are products of the construction boom of the 1980s. Speed became of the essence as contractors organized tradesmen to complete each phase of construction as quickly as possible. But framing specialists may have little incentive to push on to the higher skills of finish carpentry that would also make them more skilled as framers.

I took the salesman's recommendation, looked up John Conrad, and went to look at his recent work. It stood up well to the standards I set for myself. I liked more about John's work in progress than the quality of the frame. I noticed the care he took in managing the site. The stock was neatly stacked, and there was little trash or debris lying about at the end of the day. Quality's in the details, the saying goes, and it's true. How a tradesman cares for and stores his tools — even whether he keeps his trucks neat and clean — can tell you a lot about the quality of work he is liable to turn out.

The fact that John had gone to technical school, where he mastered the engineering aspects of framing, was reassuring to me; too many framers lack this grounding, and it's difficult to get precision when a framer doesn't understand the task technically. I asked John to submit an estimate based on Jock Gifford's plans. His bid, in my view, was a little low. It was a little less than my reckoning of the materials' cost for the frame. A good rule of thumb is that the labor cost of framing should be roughly the same as the materials' cost. My materials' cost was inflated a little because I was electing to use an engineered-joist system rather than two-by-twelve dimensional lumber to provide the understructure for the floors. The engineered joists were about $2 a foot versus the two-by-twelve planks at less than $1. As compensation, the installation of the engineered joists required less labor.

I agreed to give John the job, but I knew — and said at the outset — that we would have to adjust his price as we went along to take any unknown factors into account. For instance, I wanted the framers to remove the wood sill, which I had myself secured to the top of the foundation, install a moisture barrier under it, and reinstall it; they wouldn't have anticipated this step from studying the plans, and I hadn't anticipated it when I laid the sill. I also wanted to take into account that the framers would have to spend more time than usual moving materials on the site.

Joists had already been unloaded out by the driveway entrance — about thirty yards from the foundation — from where they would have to be hand-carried by the framers as they needed them. The deliverymen had left them there because their tractor trailer could not negotiate a curve in the new driveway that existed to spare a large pine tree. I wanted to pay the crew for the extra handling time. Fairness on my part, I figured, would engender fairness on theirs.

September 1992 had arrived cool and sunny, perfect construction weather with neither high heat nor humidity to warp any wood stacked at the site. I was more than a little anxious to get the frame up and the roof on before winter. From his first day on our project, John Conrad had Bill Delaney with him. Bill was older than John and ran his own contracting business. Depending on who lined up the job, each had frequently been the other's boss. After three days, John added a third person to his crew — Richie Woodward, a crew-cut ex-Marine he and Bill both called "Woody."

The reason they had to remove the joist sills was that I didn't want the ends of the joists to have direct contact with any concrete in the foundation. If the joists absorbed moisture from direct contact with exterior concrete, they would eventually rot. Therefore I wanted to install a moisture barrier that would go under the two-by-four sill, then up the inside and across the top of Lenny's little foot-high concrete block addition to the foundation. For the barrier I used a roofing membrane ordinarily used along the edge of a roof to prevent damage from ice damming. Only an eighth of an inch thick, it has a polyethylene surface on one side and a waxed paper coating on the other, which peels off to uncover a very sticky surface. We put the sticky surface against the concrete. The nonsticky side, where it went across the top of the bond-beam block, made a tight fit between the block and the two-by-ten wood sill that was to go on top of it. I felt sure no moisture could penetrate the foundation with this barrier in place.

John and Bill reinstalled the two-by-four sills I had once fastened down, then turned to the task of installing the two-by-ten sills along the top of the bond-beam block. Every step involved first checking to make sure the structure was both level and square. I knew that the poured foundation walls were level and square. But since I had checked Phil McCarthy's work, we had added two elements: Lenny's bond-beam block wall and the granite veneer sitting on the outside ledge of the poured concrete wall.

August 1992. The foundation is poured. The next step is installation of a granite veneer on the outside of the bond-beam block sitting atop the foundation.

I brought my builder's level to the site. Using the level, I sighted John's rule while he held it on top of the foundation. We checked every four feet around the perimeter, logging the data on a piece of paper so that we could analyze any variations we found. Phil McCarthy, Lenny Belliveau, and Roger Hopkins had done a good job; the foundation was almost perfectly level. If there had been more than a quarter of an inch variation we would have had to shim the sills with wood shingles until they were level.

A rule I adhere firmly to is: if you start square and level, the rest of the house goes together smoothly. I once got accused of building an out-of-square house. An examination of the problem quickly revealed the true culprit. I had indeed built a square house, though it took me quite a while to demonstrate that fact to the client, who seemed to want the problem to be my fault. After I had done my work, a floor man came in and proceeded to lay a tile floor two inches out of square. It really made everything else look off-kilter. Fortunately there was a way to fix the tiles without redoing the house.

The framers fastened the two-by-ten sills to the top of the bond-beam block with the anchor bolts that Lenny Belliveau had embedded in the concrete for that purpose. When they went to the stockpile to get the sills, the framers rediscovered another rule of construction: material seldom comes loaded and unloaded the way you intend to use it. Our first delivery contained two-bys for the sills and 150 sheets — two forklifts' worth — of three-quarter-inch tongue-and-groove Southern pine plywood. It was enough plywood to cover the first- and second-floor decks of the main house and the deck of the ell. Needless to say, the two-bys for the sills were on the bottom and the plywood on top. All the plywood had to be moved and restacked before John and his crew could get to the first wood they needed.

All of the two-bys in the first delivery were for sills and therefore were pressure-treated. Some people question the use of pressure-treated wood, but I don't. The process is this: the wood is dried, then impregnated under pressure with preservatives, usually chromated copper arsenic, which make the wood fibers toxic to fungi and insects, including termites. Southern yellow pine and other selected species (Western hemlock, Douglas fir, Ponderosa pine, among others) are used for pressure-treated stock. The preservatives bond well to the cell structures of those woods, making the chemicals less likely to leach out.

The chemicals used in the process give a greenish cast to the wood. Some manufacturers guarantee the treated wood won't rot for at least thirty years. It depends on the level of treatment. I do recommend checking the preservative retention ratio, which is measured in pounds per cubic foot. Although some wood intended for aboveground use is graded at 0.25, I wouldn't use any graded below 0.40, the grade at which the wood can be exposed to direct ground contact. Also, I look for the mark of the American Wood Preservers Bureau, which ensures that the treatment process used in the wood has been independently tested and approved.

A couple of cautions about pressure-treated wood. You can't safely burn scraps of it. Burning produces toxic smoke from the preservative. It is wise to wear a dust mask when cutting pressure-treated wood. Some homeowners don't want pressure-treated wood decks — to avoid any possibility that children playing or crawling on the decks will absorb preservative from contact with the wood. If that's the case, they may want to use redwood or cedar, which is naturally insect- and rot-resistant but costs more than other soft-woods.

"Better watch out," Bill Delaney said on the third day the framing crew was on the site, "might get some joists on today!" His sarcasm reflected the frustration every builder feels, that getting the job laid out takes so much time; there seems to be so little evidence of progress at first. John Conrad and I had already spent a lot of time "laying out" — marking on the sills where each of the joists was to be installed, defining openings in the floor for fireplaces and hearths, and checking where we needed doubled joists to carry extra structural loads. I had sent a set of plans to the manufacturer of the engineered joists. The company laid out the joist system I needed and sent the necessary material with a plan of its own, a cut list, and all of the instructions needed for installation.

Each of the engineered joists is a wooden I-beam. The top and bottom chords — the crosspieces of the "I" — of the kind we used are made of plywood. You can order chords of different widths depending on the load you want the joist to carry. Most of the joists we used have inch-and-a-half-wide chords, but for the longer spans in the floor of the ell we used two-and-a-quarter-inch chords, and in the longest span, in my office, we used three-and-a-half-inch chords. The web — the stem of the "I" — is made of oriented strandboard (compressed, glued wood fiber) three-eighths of an inch thick. The web is glued into a groove in each of the chords. Manufacturers make this product in very long lengths, and the user cuts it to fit.

The structural advantages of engineered joists over ordinary wood planks are that the engineered joists are more uniform in measurement and not subject to shrinking or "checking" (cracking) over time. The conservation value is that the joists use less wood fiber to provide more strength than equivalent planks use. As is standard in residential construction, the joists in our house were spaced "sixteen inches on center," meaning that it is sixteen inches from the center of one installed joist to the center of the next one. This standardization allows easy use of such products as plywood and insulation; for example, a sheet of plywood is ninety-six inches long, which is an increment of sixteen; therefore, both ends will fall on the center of a joist. Fiberglass batts are fifteen inches wide, which means they will fit snugly between joists without having to be cut.

The ell is narrow enough that the floor joists could run from the north wall to the south wall without any intermediate support midway. But in the main house the span is too long in either direction to have unsupported joists. The framers therefore installed a center beam in the basement running north to south at the proper height so that each of the east-west–running joists could rest on it. They made the beam by nailing together three nine-inch-high LVLs (laminated veneer lumber), which are an engineered wood product. They consist of thin laminations of wood glued together like plywood except that the finished product is thicker than standard plywoods — one and three-quarter inches thick in our case. Later on in the framing, John Conrad installed LVLs wherever we had to support large loads over long spans, for example, as a header over the garage doors.

As each joist was put in place, it was blocked at each end and over the center beam so that it couldn't tip sideways. Engineered joists have tremendous vertical strength when upright, but lying on their sides they are as flexible as one-by-three strapping. This flexibility could be interpreted as flimsiness if one didn't understand the engineering principles embodied in their design. Bill Delaney seemed more skeptical of the joists than John Conrad was. "It supposedly won't shrink," Bill scoffed one morning, looking at one of them. "It's freeze-dried."

John, as the subcontractor in charge of the framing, admitted to me when we could see the job running smoothly, that he was nervous when I gave him the job. A manufacturer's plans — even blueprints created by a designer as meticulous and thoughtful as Jock Gifford — can be incomplete or contain small errors. So good framers have to be flexible and observant.

The framers were wearing the usual "badges of courage." Bill Delaney was sporting a nasty-looking scab on the top of his head, where he'd rammed into the center beam of the main house. He hadn't even noticed it was bleeding at the time, and kept working while blood trickled past his ear. If it didn't bother him, it didn't bother me. But Laura, who happened to be visiting the site at the time, didn't appreciate that I continued to give Bill some suggestions about framing before getting him some first aid.

*Bill Delaney looks
ready to meditate
perched confidently
on an engineered joist.
He waits and watches
while John Conrad
installs a joist hanger.*

"Framers are the only subs [subcontractors] with personality," John insisted during a coffee break. "We work in the sun and we've got the tools. Everybody wants to work with us in the summer." But not in January, he might have added. Framers, John thought, spend the most time on a site and earn the least. Some crews, when the New England economy took a nose-dive, were making as little as $1.75 per square foot. I wouldn't have accepted a bid that low to frame our house. Whatever dollars one thought one was saving would be lost in cut corners, hasty work, hidden mistakes.

I was trying my best not to make any changes in the plans as we went along. Changes cost money — in both labor and materials — and can quickly drive a budget out of sight. "People will look at a room I've just framed," said John, "and say, 'Oh, it's too small.' But I tell them, 'Once you build a wall and stand it in place, there's an extra charge to alter it.' You give them too much freedom, and they'll take advantage."

In the division of labor for installing the deck, Woody, the youngest of the three but already with flecks of very premature gray in his hair, got the grunt work. He lifted two sheets of three-quarter-inch-thick plywood at a time from a stack at the edge of the driveway and carried them twenty feet to the edge of the foundation and then to the place where John and Bill were waiting to fasten them down. He could make eight straight trips — a half ton's worth — without a murmur of complaint.

"My father always said," Woody told us, " 'Take two steps at a time, so when you're old, one won't be a problem.' I take two sheets, so when I'm as old as Bill, one won't be a problem."

John and Bill squirted adhesive on top of the joists where the plywood would rest. Adhesive is an important ally in the cause of eliminating squeaky floors. When plywood isn't fastened down tightly, every step taken on it forces it down on the joist, and every step lifted from it allows the plywood to spring back to where it had been — the prescription for a squeaky floor. Sheets moving up and down the shanks of less than fully driven nails can make every tread an audible irritant. That's why I always use construction adhesive as well as nails for attaching deck sheathing. Two systems greatly reduce the chance that you will hear unwelcome music later from your sub-flooring. The gluing takes time, but it's worth it.

Midpoint of installing the first-floor deck. The main house deck in the background has been sheathed. The framers are sheathing the ell. In the foreground, the frame of the garage wall can be seen as preparation for the installation of the deck of the office/garage wing.

Eventually the deck would have two layers of plywood in order to make a base to support an inch-and-a-half layer of gypsum concrete in which we would embed the tubing for radiant floor heating. But for the time being we were installing only the first layer — of three-quarter-inch, tongue-and-groove plywood. Construction plywood is another building material that has been around so long we tend to take it for granted. What is constantly improving is the glue holding the plies together. There's an argument to be made for not cutting corners on plywood. Cheap plywood means cheap glue, and delamination down the road.

Three-quarter-inch plywood is twice as strong as half-inch, makes a better subfloor, and comes tongue-and-grooved along the long edges; the connection keeps the sheets from shifting where they span from one joist to another. The shorter ends have no tongue and groove because the entire width is fully supported by the joists into which they are nailed. One warning: plywood exposed to moisture will swell. It's good practice to leave space, maybe an eighth of an inch, between sheets used on any subfloor exposed for a time to outdoor weather conditions. With too tight a fit, during a long rainstorm, the plywood could swell enough over the length and width of a deck to buckle the sheets.

Staggering the sheets of plywood so that there weren't any adjacent seams on the same joist, John and Bill used a piece of two-by-four to tap the long tongued edge of one sheet into the grooved side of its neighbor. John secured each sheet with two or three hand-driven nails. Bill followed along with a nail gun, shooting nails every four to six inches where the sheet rested on a joist.

The nails were the first of what might eventually total eight hundred or nine hundred pounds. A rule of thumb is three hundred pounds of nails per seventeen hundred square feet of house, and, if anything, I used well over the average. For better holding power we were using nails coated with a resin. Friction between the nail and the wood as the nail is driven temporarily melts the resin, which bonds the nail to the wood as it cools again. John was hand-driving all his nails, which surprised me in this high-tech world. I use a nail gun whenever I can. The nails come in rolls, or clips, and are driven by an air compressor–powered piston into the wood. With a gun, the nail passes through the wood with little disturbance.

On their off time, I supposed the framers were telling their families anecdotes about working on Norm and Laura's house. Anecdotes such as an account of the Friday Laura arrived at lunchtime to give them their first check. After telling them that she still "couldn't see the house," she admitted that she was worried we might have started a never-ending project.

"I'm nervous," she said, as they all sat around a picnic table like one I had made a couple of years ago for a *New Yankee Workshop* show.

"Really?" asked Bill. "How does the house look to you now? What do you see?"

"What I see here," said Laura, "is the biggest unfinished project Norm has ever started!"

John and Bill were beside themselves. Bill, especially, couldn't wait to get home. "That's exactly what my wife says about me," he grinned. "Wait till I tell her Norm Abram never finishes anything either!"

Arrivederci, Exercise Room

NONE OF LAURA'S and my meetings with our architectural designer, Jock Gifford, ever took place at our old house. We always met him at his office, or at the new property, or at some other convenient place. If Jock had visited our old house, I'm sure one of his first questions would have been: "Where are you going to house your birds?"

Laura has had a lifelong interest in tropical birds. When she was a child, her family frequently kept a caged parakeet called a "budgie" (short for Budgerigar) as a family pet. On family vacation trips, Laura stopped in every pet shop she could find to see the tropical birds, especially the parrots. As she got older she began reading books to became familiar with their ways and the proper care of them.

A few months before we began talking about a new home, Laura bought her first parrot. I was with her when she went into a pet shop in Marlboro, Massachusetts, and found a parrot who came readily out of its cage and sat on her shoulder. Laura was sure this parrot was meant for her, and I agreed that we should buy him, Goofy by name, and a cage and take him home.

Goofy didn't last very long at our house, only a couple of months. Laura came to think of him as the parrot from Hell. He bit everyone he could; and he made a lot of noise, especially in early morning hours close to sunrise when I'm getting my best sleep. Laura took him back to the pet shop in

Marlboro and told the proprietor she guessed it was her inexperience that made living with Goofy such a disaster. She wasn't ever going to have a parrot again. I didn't try to dissuade her. I began to sleep better around sunrise.

But a few months after Goofy left the house, Laura visited another pet shop, and there she saw and fell for a baby Senegal whom she later named Gus, who is still part of our household. Right away she learned one of the major differences among tropical birds. Goofy was an imported bird. There is quite a large trade in such birds. Traders knock down trees in the tropics to raid nests and take the birds when they are still babies; but they are always then partly wild, too.

Gus, however, had been bred domestically by a dealer in birds. Very soon after hatching — as early as the tenth to twelfth day, before their eyes are open — domestically bred birds are taken from the nest and hand-fed by humans. The baby birds then bond with the humans who feed them rather than with their parents.

Laura mastered the technique of hand-feeding baby parrots. She taught Lindsey how to do it. She even taught me well enough that I can hand-feed a baby parrot when no one more experienced is around to do it. Some dealers who didn't have time to do all of the hand-feeding themselves began to call Laura. One would call and say, "I've got two babies ready to come out of the nest. Why don't you take them on consignment?" So Laura began bringing baby parrots home, caring for them until they were mature enough not to need hand-feeding and could be sold to new owners. Occasionally one of these babies worked its way so deeply into Laura's affection that it didn't ever get returned to the consignor.

Our largest parrot is a macaw named Blue. He is a very handsome bird with bright blue back and tail feathers and equally bright orange-yellow breast feathers. Sitting on a perch, he measures about a foot from head to claw. His tail extends another ten inches or so. His wingspan is about two feet, although his wings are clipped to inhibit his flying. His beak is powerful enough to enable him to snap a broomstick in two if he is of a mind to do so.

The first time I approached Blue, he got my finger in a pressure bite. Let me tell you, that hurts! I pulled back, once I got him off my finger, and then I was wary of approaching him again. If he moved his head toward me, I'd instinctively pull back. "That's exactly what he wants you do do," Laura said. "He's got your number. If you don't approach him without hesitation, he'll always think he's dominant over you." I took her advice about Blue, and the two of us get along well now. But I hope I never get pressure-bitten again.

I can remember a little of what it was like when I first got seriously interested in making furniture with power tools. I couldn't get enough of it. When I wasn't in the shop, I was thinking about what I was going to make when I got to the shop. Laura could probably tell you more than I can of how preoccupied I was with my new interest. Even today I'm late getting home lots of nights because I can't pull myself away from a woodworking project I've started. Caring for beautiful tropical birds seized Laura's interest the way woodworking had earlier seized mine. Of course, I wasn't nearly as interested in birds as in woodworking. Since Laura and I were getting into the process of finding another house, and then of building a new house, I didn't think she had found this new interest at the best possible time. The schedule of our household seemed to revolve more around the birds than around the Abram family.

The day before Easter of 1992, Laura came home carrying a box. I was in my office working at my computer. She set the box down and said it was a surprise for me. "Well, it better not have feathers," I muttered. She opened one end of the box and an African gray parrot peeked out. "Hello," he said. Then he hopped up on my arm, looked at me, looked at the computer. "Work, work, work," he said. Boy, this is my kind of bird, I thought.

It was Peter who first got me to think of the parrots as our extended family. For whatever reason, he became "my" bird. Peter has a vocabulary of about two hundred words, phrases, and sounds. He tends to pick up things said with emphasis, so one has to be a little careful what one says around him. He can imitate the ring of the telephone or the sound of the fax machine so accurately that I almost start toward the instrument. He learned my credit card number because Laura used the telephone near his cage when she ordered merchandise using my card number. Peter would confuse the person taking the order by repeating the number audibly as Laura recited it, and sometimes he would terminate the call prematurely by saying "Bye, bye" in Laura's voice. She soon learned to place her orders using another phone.

When Peter is out of his cage and wants to get my attention, he plucks a single hair out of my beard; it *does* get my attention.

Peter is one of those birds born in the wild and then domesticated. He has never been happy close to our other parrots. He won't speak at all when other birds are close to him. I think he will live in the family room in the new house. Blue, the large macaw, will probably have his big white cage in the all-seasons room. But the remaining several parrots of various kinds — Hawkhead, Amazon, Cockatoo, Lovebird, Senegal — need a room of their own.

The obvious place is the extra room on the same story as my office in the back wing. As an exercise room, it was going to be one of the least used rooms in the house even if I exercised every day, which is unlikely with my schedule. There is plenty of room for exercise equipment in the full basement under the ell. I can also put exercise equipment in one corner of my shop once I get it built.

In our house, there is always a new parrot story to relate. One morning a few months ago, to cite just one example, when Laura came downstairs, Peter made a sound she hadn't heard before. It sounded like someone clearing his throat. Peter made the same sound four or five mornings in a row.

Finally, a theory about the new sound occurred to Laura. "Norm," she said to me, "have you been clearing your throat and spitting in the kitchen sink in the morning?"

"No way," I said, wondering what led to such a question.

"I think you have."

"What makes you think that?"

"Peter told me. Norm, don't spit in the kitchen sink."

And I haven't. But the sound is in Peter's repertory now. Once in a while he makes that sound, and Laura gives me this funny look. Is Peter putting me on? What can I say?

*My attitude toward
Laura's parrots was
hesitant until
Peter, an African
gray, captured my
interest with his
vocabulary and
imitation of household
sounds.*

A Frame-Up

BY LABOR DAY of 1992, the first-floor deck was complete. Usually the completion of the deck assumes the completion of other phases of construction: the foundation has been backfilled, the site rough-graded to provide level areas around the foundation for storage and work stations, the septic system installed, and the electrical service operational at least as far as the main panel board. My dream house wasn't matching the norm (no pun intended). The electrical contractor hadn't yet pulled the wires through the conduits I had buried in a trench in August; the septic system existed only on paper; and there wasn't much grading near the foundation because piles of various materials cluttered the area.

I had seen enough of the framers' work by this time to know their style as individuals and as a team. Like framers everywhere, they wore as little clothing as weather permitted. Early September was blessing me with some good hot workdays (and distressing me with some rainy days), so all three were shirtless much of the time above their shorts or jeans. As the days cooled, they added more clothing. John Conrad wore a soft cloth apron tied about his waist to hold his nails. His hammer and measuring tape sat in separate holders attached to his belt. Woody Woodward and Bill Delaney didn't like to carry nail aprons centered in their laps; every time they crouched to work, they found the nails in their way. They had switched to soft leather pouches that rested on one hip and held, in addition to nails, their basic hand tools — hammer, measuring tape, and carpenter's square. Because John's cutoff sweatpants lacked pockets, he tucked a carpenter's pencil into each of his ankle-high socks. "If I'm up on a ladder or wall and drop one," he said, "I don't want to have to climb down right away to get it."

Thinking ahead to minimize climbing and lifting is characteristic of a good framing crew. Another of their practices is to take time to remove every nail from discarded pieces of wood. Protruding nails can cause injuries or damage tools. "My first boss," said Bill, "would send us home if he caught us bending a useless nail over. He'd say, 'I can't afford to lose a day's work because of sloppiness.' " "There are two rules," Woody added, "and that's one of them. The other rule is: once you pick something up, don't lay it down before using it." Bill had lasted a long time as a framer, Woody observed, by following the second rule religiously.

Before the framers started to raise walls, I reviewed the plans again with them, looking at "floor plans," "elevations," and "sections" for the main house. Floor plans show the layout and dimensions, including the centerlines of all windows and doors, of all the rooms; the perspective is as though you are looking directly down on the rooms from above. Exterior elevation drawings show a facade and include both finished floor elevations and plate elevations for the tops of walls; the windows and doors are keyed to a list of sizes and specifications elsewhere in the plans. An elevation drawing shows the equivalent of what you get by setting up a camera and taking a picture of something; it doesn't show what's behind anything. A section drawing, on the other hand, is like what you see when you cut through a sandwich, revealing its interior layers. A section can be a vertical cut through a whole house, showing how the floor, walls, and roof come together, or it can be a detailed section — for example, a wall section, showing every component of the wall, from the footing to the eave line, including insulation, floor, wall, and eave details with critical dimensions. Equally, a section can be a horizontal cut either of the whole building or of a detailed part of it. Framers have to consult all three kinds of drawings, and integrate them mentally, in order to build a frame accurately.

Most framers these days build the frame for a wall flat on the deck, and then lift and push it up into its proper vertical position. This procedure is much easier than nailing everything together in midair. In my own work as a carpenter and contractor, I have often raised walls with wall jacks. Two men slowly cranking jacks can raise a seventy-foot-long wall frame, a task they could never accomplish manually.

John Conrad's crew did most of their raising by muscle power, but when they finally got to construction of the largest walls of the office wing, they brought in jacks. As they built each wall on the deck, they secured its bottom plate to the deck with short pieces of "banding" (metal straps that secure lifts of lumber or plywood together for shipping). The banding acts as a hinge. The wall pivots on the banding as it is raised yet is restrained so it can't slide off the deck. Care must be taken as the wall approaches a vertical position; the wall has to be secured immediately with temporary bracing to prevent it from falling over away from the deck, or back onto the deck where it was built.

On a windy day, raising a wall can be a hazardous adventure, especially if the plywood sheathing has already been applied to the studs. Then the wall becomes in effect a large sail as it's lifted. Woody, I learned, had been injured once when he, Bill Delaney, and a third carpenter tried to raise a wall at the end of a day. The inadequately secured bottom of the wall pulled away from the deck and the whole wall slipped and slid. Woody's head — turned sideways — was caught between the wall and the floor. Luckily, he suffered only the loss of a few teeth.

To begin framing the exterior walls of the main house, John and Bill first snapped chalk lines on the deck around the entire perimeter to show the inside line of where the frame should stand. Then, for each wall as they came to it, they placed a row of two-by-fours on edge along the chalk line and a second row on edge right next to the first row. The two-by-fours nearest the chalk line became the sill plate on which the studs sit; the other row of two-by-fours became the top plate to which the top ends of the studs are attached. John laid the rows next to each other temporarily for convenience. The only difference between John's layout procedure and mine is that I always do my layouts on the broad sides of the two-by-fours, not the edges. Both methods accomplish the same purpose; it's a matter of personal preference.

On both rows John carefully measured and marked all the openings in the wall. They are called "rough" openings, but there's nothing rough about them. They determine the exact placement of windows and doors. The specifications of each one must be known before the wall is framed. Windows of the same style and quoted size made by different manufacturers may have small but critical differences that will affect the dimensions of the rough openings.

Each rough opening in a load-bearing wall has a "header" at the top — a strong wood member that supports the weight of the floor or roof above it; the header can be made by doubling standard dimensional lumber or by using LVLs specified by an engineer. The ends of a header are supported by studs called "jacks," which transfer the header's load down to the deck. On non–load-bearing walls, a couple of two-by-fours on the flat should be sufficient as headers. Window openings have horizontal members across the bottom called "rough sills" on which the window unit will sit. Generally there are some additional short studs called "cripples" between the rough sill and the sill plate attached to the deck.

When John had marked both rows of two-by-fours for rough openings, he measured and marked them for the location of the studs spaced sixteen inches on center, and for the shorter cripples between windows and sill plate, and between doorways and top plates. With two sets of two-by-fours bearing identical markings, John separated the rows on the deck by the height of the studs, so that the second row could become the top plate for the wall; he and his cohorts filled in the marked frame piece by piece. The final installation on each wall was to double the two-by-fours of the top plate; the floor joists for the second floor would eventually sit on the upper layer of the plate.

After the framers had nailed the components of a wall together into one unit and aligned the frame perfectly along the chalk line where the sill plate was supposed to sit, they attached the metal bands to the sill plate and deck to keep the wall from sliding when they raised it. They checked the wall once again for squareness by measuring the diagonals to make sure they were equal. Then they applied half-inch CDX plywood sheathing over the studs, making sure it was cut flush with the edges of all openings. (CDX plywood is C grade on one side, D grade on the other; it is bonded with adhesive that will tolerate eXterior use.) The wall, as they raised it, was thus already sheathed.

Many houses under construction these days have exterior wall frames made of two-by-sixes rather than two-by-fours. The thicker sixes allow the builder more room between the studs for insulation. If sixes are used, the studs can be spaced twenty-four inches on center rather than sixteen inches as is customary with two-by-fours. Since each stud interrupts the insulation barrier, leaving cracks where air can permeate either in or out, the fewer studs the better.

I elected to go against this trend and have the exterior walls built of two-by-fours. In the space gained from using thinner studs, I planned to install, on the inside surface of the studs, a solid blanket of inch-thick, foil-faced, rigid polyisocyanurate insulation. Building codes in each area or state stipulate the minimum amount of insulation required of a builder. Insulating capacity is measured by the resistance to conductive heat flow — referred to as its R value — of a given material. In the summer, insulation keeps warmer exterior air from invading a cooler house, and in winter it keeps precious warm air inside from escaping to the outside. The higher the R value, the lower the conductive heat gain or loss.

If I had used two-by-sixes for the frame and put six-inch-thick batts of insulation between them, I would have achieved R-19 insulating capacity, more than enough to meet Massachusetts code. The unfaced batts I put between the thinner studs have R-13 value. The rigid insulation added another R-7.2 for a total of R-20.2. The greater advantage, however, was not the higher R number; it was having a continuous barrier of leak-proof solid insulation, uninterrupted by the studs. Also, the aluminum foil moisture barrier on the rigid insulation is more effective than a polyethylene moisture barrier used with unfaced batt insulation.

After the exterior walls were erected and the corners nailed together and made plumb, the framing crew nailed small blocks of wood of equal thickness to the outside top of every wall end. They hammered another nail partway into each block and ran a string tightly from one nail to the other, tightly against each block. Since the string was tied the same distance from the frame at each end, the framers could walk along a wall, using as a gauge a block of wood of the same thickness as the ones already nailed on, to see if

The framers have raised the frame of my office's south wall off the deck to their head height using wall jacks, and then pushed it up to nearly plumb manually with a two-by-four. John Conrad adjusts the two-by-four, Woody Woodward checks for plumbness, and Bill Delaney gets ready to nail a side brace.

there were any places where the frame was closer to or farther from the string than it should be — in short, crooked rather than straight. The technique is very simple, but it works.

The framers dealt with any straightness problems and made each wall more secure by installing temporary braces called "springboards" every six to eight feet. Made of one-by-eight spruce, the twelve-foot springboards are angled from the top plate down to the deck and secured with nails at both ends. For adjustment purposes, the framers set one end of a four-foot piece of one-by-eight on the deck so that its other end could be wedged up against the underside of the springboard. Where the wall is out too far, the shorter piece is wedged just enough to bend the springboard and pull the wall in until the gauge block fits perfectly between the outside top of the frame and the string — at which point, the short board is nailed to the springboard and deck. The springboards keep the walls straight until another floor or a roof is framed and sheathed above them. John Conrad used the same springboard technique for the interior bearing walls to hold them straight and true while additional framing was added on top of them.

At each outside corner of the house frame there is a corner post. Most framers nail three two-by-fours together to make the post where the plywood sheathing from two walls comes together on the outside and the two adjacent drywall or plaster surfaces come together on the inside. Although it is a kind of throwback to an older method of construction, I chose solid four-by-six fir posts for the corners.

"Four-by-sixes for corner posts?" queried Woody. "That went out with high-button shoes." Still, John showed some appreciation of the old ways by installing "windbracing" at the corners. The crew let in two-by-fours into the studs at an angle from the top of the corner post down to the sill plate. Windbracing was common when boards rather than plywood were employed to sheathe the frame. I was pleased to see the crew add them as an extra measure to keep the house from racking in a high wind. I also knew that our town's building inspector requires windbracing even when the sheathing is plywood.

Most framers concentrate on exterior walls and interior bearing walls until the exterior frame is up and the roof is on. Then, with protection from the roof for interior work even on rainy days, they turn to the remaining interior or "partition" walls. John Conrad elected to complete the interior framing of the first floor of the main house before moving up to the second floor; he and his crew then installed the second-floor joists and deck to use as a platform for building and raising the exterior and interior bearing walls of the second floor.

By noon of September 10, all of the exterior and interior walls of the main house's first floor stood in place. I was immensely heartened. Surely John Conrad would have the second floor, attic, and roof framed before the end of September, and an as yet unlocated roofer would have shingles laid before first snowfall.

I fully expected the framers to rig some scaffolding around the perimeter of the house and along the center interior bearing wall when they started to frame the second-floor deck. It would take only about two hours to set it in place. Not only would the subsequent lifting of the engineered joists and plywood decking go faster, but the work would be safer. A good investment of time. Scaffolding was not in John's mind, however, which explains his strategy of completing all of the first-floor interior as a base for the second-floor deck before framing the exterior walls of the second floor. He didn't own very much scaffolding and hadn't included renting more in his bid calculations. He told me his scaffolding was a ladder and a plank. The framers planned to balance on top of the first-floor walls while measuring, marking, and fastening the engineered joists for the second floor. "It's like walking a tightrope," John said, as he inched his way along the plate on top of one wall, one pencil still in his sock and the other one behind an ear for accessibility.

Bill Delaney seemed disappointed that I hadn't used two-by-sixes for the first-floor studs and top plate, which would have given him a slightly wider catwalk. As the framers proceeded with the second-floor joists, I noticed that Bill wasn't doing much catwalking. "He says he'll be too old in two years to do any of it," laughed Woody. "I'm too old *now*," Bill lamented.

Since much of my television work stresses the wisdom of using the right tools and equipment for a safe and efficient job, I found myself a little baffled that I had hired framers whose craftsmanship was apparent but who had a basically low-tech approach to their craft. I even had to lend John my nail gun. "By the time you roll out the compressor and hoses," John said, defending his preference, "you can have the wall nailed by hand." To each his own, I guess.

John and Bill studied the second-floor plan and compared the amount of joist stock needed with what remained piled on the site. They realized they had to do some cutting. On the plans, the engineered joists were specified at thirty feet even, but as delivered from the lumberyard each was thirty feet six inches long. Once the joists were cut to exact length using a plywood jig to save measuring each piece with a tape, Bill and Woody carried them one by one up a ladder to where John was perched on top of the first-floor wall to receive them. The flexibility of the joists makes them trickier to handle than ordinary dimensional lumber. In fact, the manufacturer recommends that the joists be kept on their edges so as not to damage them. After three were stacked on their sides atop each other, Bill yelled to John, "Can't you walk on them yet?"

"I'm not gonna," said John.

"We can walk on them down here," teased Bill from where he and Woody were clambering over a large stack of joists.

"Yeah," John replied, "because they're stacked ten deep. When we get all of them up here, I'll be able to walk on them, too." It wasn't long before John and Woody, perched on the top plates at opposite ends of the second floor, nailed the first second-floor joist into place. "This part's going easier than the first floor," said Bill approvingly from below. Indeed, they did seem to have mastered the technique of handling engineered joists.

Although it seemed to rain every other day that September, the frame rose at a reassuring pace. In time that I carved out of my schedule, while the framers pushed forward on the second floor, I did a little carpentry. The two-by-twelve lumber for the stair "stringers" (the inclined notched boards that support the ends of the treads) had been delivered with the framing lumber. I did the layout for the two partial flights of stairs in the office wing, cut the stringers, and installed them with temporary treads made from three-quarter-inch plywood. But I couldn't yet set the stair from the first to the second floor of the main house in its final position. We intended that stairway to curve up along a wall behind which plumbing and vent pipes were to be hidden. The plumbing wasn't installed yet, and the wall itself couldn't be finished. So I constructed a temporary stair rising straight up on the other side of the hallway where the framing was complete. Now everyone could get from one level of the house to another without climbing ladders.

With the second-floor exterior and interior bearing walls in place, straightened, and braced, it was time to start framing the roof. The attic was not meant to be finished space, so we framed its floor with standard two-by-tens instead of more expensive engineered joists. The floor would be plenty strong for storage space and mechanical equipment; the joists were deep enough to take nine inches of required fiberglass insulation between them. Only five openings had to be framed in the attic — two for chimneys to pass through, one for the base of the large skylight shaft over the second-floor hall, one for an attic pull-down stair in the walk-in closet off the master bathroom, and one for a skylight in the bathroom off the guest bedroom.

I hadn't stopped keeping detailed records of everything associated with the house, but I had stopped adding up costs to get running totals. Something told me that I would get nervous and enjoy the realization of my dream less if I totaled expenses every week. Still, I looked for ways to save money without cutting quality in any significant way. For example, I had the framers use five-eighth-inch CDX plywood instead of three-quarter-inch tongue-and-groove plywood for the attic floor — and then only in the area that was easily accessible without banging heads on roof rafters. I suggested to John that his crew carry plywood to the attic but install only what they needed to make a work platform to frame the gable ends and the roof. Much later, after the framers had moved on to other jobs, my son Bobby cut the rest of the attic floor plywood to size and nailed it in place. (Bobby is my stepson, but he thinks of me as his father and I think of him as a son. So I shall hereafter refer to him as my son.)

Using the partial attic deck as a platform, the framers laid out and built the gable ends of the main house. The task was similar to framing a wall for a lower story, except that the gable is triangular rather than rectangular in shape, and is topped by an end rafter rather than a two-by-four top plate. The framers sheathed the gables as they lay on the deck, then raised them into place and braced them. From the peak of one gable to the peak of the other

Standing in the ell, I am pointing, for the benefit of an unseen framer, to the temporary stairs I built from the center hall of the main house to the second floor. The final stairway will be on the opposite side of the hall.

they set a two-by-twelve ridge beam, from which they ran two-by-ten roof rafters from the ridge beam down to the edge of the attic deck. The main house frame was then complete.

The office wing at the back of the house was their next target. The deck for that wing was already installed; the framers had built it at the same time they built the decks of the ell and the first floor of the main house.

While the frame of the main house presents very few variations from the frame of a classic colonial box, the back wing is one variation after another. The deck itself is on two levels joined by a short partial stairway. The mud hall and pantry/laundry room deck is on the same level as the adjacent great room of the ell; but the deck for my office, office bathroom, and the bird room (formerly my exercise room) is three feet higher. Consequently, the only conventional elevation in the office wing is the south gable wall above the garage doors. Inside the gable wall is my office, with side walls only five feet high but with a cathedral ceiling that rises to thirteen feet at the peak. John chose to build this gable wall first, laying it out on the deck, assembling it, sheathing it, and raising it into place.

The north gable wall is not symmetrical because it borders two different floor levels — the bird room on the higher level and, three feet lower, the pantry. It is a little more complicated to build a wall in one unit in that circumstance. Yet it is more practical to do so, since the point at which the floor level changes is not the midpoint of the wall, directly under the peak of the gable, which is the only reasonable place to divide a gable wall frame. John built the wall as a unit but postponed framing the pantry window and installing some of the exterior plywood sheathing until he had the rest of the wall raised — a good compromise.

John faced a similar asymmetrical situation when framing the east wall of the office wing. The north half of the long wall sits at pantry level; the south half is three feet higher at the level of my office and bathroom. But whereas the gable wall could be broken into two sections easily only where the rafters meet, the east wall could be broken into different sections anywhere at all. The most logical place to make the break was where the floor level changes.

Since the details of the eyebrow-like window in the office portion of this wall had not been drawn — all I knew at this point was that I meant to have such a window — I had John frame and sheathe the wall as though it didn't have a window in it at all. I thought it easier to cut out a few studs and remove a little sheathing later when I framed the roof over the window myself.

Probably the easiest wall to frame in the entire house was the west wall of the office wing. It was only five feet high and it had no openings at all!

With the back-wing walls raised, straightened, and braced, the framers moved up to its roof. It has a very different structure from the roof of the main house. The main house ridge beam is of two-by-twelve dimensional lumber. The ridge beam of the office wing is made of two one-and-three-quarter-by-sixteen-inch LVLs nailed together. It is supported at midpoint by a six-by-six fir post that is directly above a concrete-filled metal column in the garage. The reason for the heavy-duty beam and midpoint post is that the ridge beam is forty-four feet long. Since my office and the bird room have cathedral ceilings, there was no opportunity to add collar ties to transfer some of the roof weight; much of it has to be absorbed by that laminated ridge beam.

Knowing the weight of the ridge beam and how difficult it would be to hoist it up ladders, and knowing that the framing of the rafters in the cathedral-ceilinged office wing meant working at greater heights than the framing of the main house involved, I offered the framers the use of some scaffolding I had access to — steel pipe frames with aluminum platforms. They readily accepted the loan. If I kept up these enticements, there's no telling how soon the framers might go high-tech on me.

John had three skylight openings to frame in the roof of the office wing — one in my office, one in the bird room, and one above the stair leading from the mud hall up to the office level. Because all of the details had not been worked out for the roof above the eyebrow-like window, I had John frame an oversize opening where the roof could be framed in a curve over the window and blended into the surrounding roof.

As the framers worked on the back wing in late September, the season began to change dramatically. The weather report one morning forecast snow in the mountains of northern New England and overnight temperatures, where we were, down into the thirties. John and his crew started work in the mornings wearing jeans and flannel shirts, then stripped to their waists in warm afternoons. Between hammer blows they filled the air with macho chatter — women and sports. When John missed a nail with three separate strokes, he groaned, "I feel like the Red Sox," referring to the locals famous for their late-season dives into ineptitude and beloved despite them.

My timetable was slipping deeper into the fall. I no longer felt as confident as I had in August about getting the house completely framed and roofed before winter. The timbers Tedd Benson was drying for my trusses had twisted in the kiln, he reported — someone had set the thermostat improperly — and the date for the timber-frame installation had been set back at least to mid-November.

I decided, in light of the postponement of timberframing, not to have John Conrad's crew build the north wall of the ell only to leave it lying on the deck for several weeks until the timber frame arrived. They moved back to the main house to finish framing the interior walls on the second floor and the skylight shaft above the second-floor hallway. Their craftsmanship was of such quality that although I had sketched the shaft on a staircase stringer — without dimensions and with only a rough perspective — they proceeded quickly and smoothly to execute it the way I envisioned it.

I walked methodically through the entire house in late September making a "punch list" for the framers. A punch list reports things that are missing or incorrectly built. Happily the list was short. It consisted mostly of notes on missed blocking or nailing for the rigid insulation and drywall. I walked through the house a second time, this time with John, reviewing my list. Normally, framers leave some interior work for a later date and move on to another job while the weather still makes exterior framing feasible; but I was eager to get everything possible done before losing their main attention to another site. All in all, we had gotten through the framing stage with minimal reframing based on afterthoughts.

Jock Gifford walked through the framed house with me shortly before the framers finished. "All along," said Jock as we walked into the pantry/laundry room in the back wing, "we've worried about this room being big enough. Sometimes laundry and pantry space get squeezed too much." He stared at the stud walls. "I think the room is going to be large enough." A few minutes later, standing in the frame of my fifteen-by-twenty-four-foot office, which seemed all the larger with its cathedral ceiling and one corner opening out into a wet bar area, Jock wondered if I hadn't overdone it a bit.

"It's a big area for a telephone," quipped John Conrad, who was accompanying us on the tour.

"Yeah," laughed Jock. "I think there'll be enough room for the fax machine, too."

October 6 was the framers' last day on the frame, though they were to return at various times to assist in roofing, timberframing, and window installation. I had been counting on the framing giving Laura the sense she had been waiting for, so far in vain, of what the finished house would be like, how spacious the rooms were compared with those in our old house. And it did. "I can relate to this," Laura said in mid-September, when the frame of the main house was almost completed. She was also beginning to take pleasure in the site. A chipmunk darted in and out of one of the stone walls as we stood looking at the frame. Laura watched it with delight. She remembered that when she was a child vacationing in Maine, there were chipmunks every-where. But at home, these days, we seldom see one, and Laura had begun to refer to them as an endangered species.

She wandered off, looking again at the plants growing wild up on the knoll. Lindsey and I set about cleaning up debris left by passers-through. The paths crisscrossing the property had suggested to us during our first wintry visit that our property was a regularly traveled route for hikers and bikers, a pat-tern not likely to change until we moved in and fully established the site as *our* territory.

Lindsey and I also swatted mosquitoes frequently as we cleaned up. Sep-tember's warmth had been ideal for maintaining the mosquito population, something we would always have to deal with because of the surrounding wetlands. One of Herb Brockert's employees had come by a couple of days earlier to remove Herb's last remaining piece of equipment, a small bull-dozer. Bill Delaney asked him to smooth out a section of driveway near the garage before he took it away. The area was a rutted swamp of mud puddles. The driver obliged Bill by scraping the drive for several minutes, but it re-mained wet and muddy.

Little by little, the site was getting more organized. Roger Cook, who had handled the clearing of the site for me as a first step in the construction, moved the logs left from the spring clearing to a more convenient place away from the construction, backfilled the utilities trench, and carted off the pile of scrap lumber to a facility that grinds it up and recycles it blended with bark mulch. Progress on the ground at this juncture, however, didn't count as much as progress on the roof. I needed shingles, a truckload of them, and an experienced roofer to install them.

Up on the Roof

THERE IS NOTHING mysterious about a gable roof; with its ridge raised parallel to the longer walls of a rectangular house, creating two surfaces of roof sloping down from ridge to eaves, it is the simplest roof design to frame and build. The term "gable" refers to the triangular section of end wall from the eave or cornice level up to the ridge. When there is an attic behind the gable, no other type of roof can be more effectively ventilated. If there are any technical aspects to worry about in a gable roof, they have to do with construction around any openings in the roof. In the case of our main house, there were going to be five openings — two for chimneys and three for skylights. We wanted these openings to be as weather-tight as the roof shingles themselves.

The gable is centuries old, so basic to American colonial design that it scarcely merits comment unless you build a house, like the famous one in Salem, Massachusetts, with seven of them. Laura's and my house was going to have four gables: one at each end of the main house, and one at each end of the office wing.

In the design stage Laura and I had to elect a roofing material. In our location and with colonial design, the choice was among slate, wood shingles, and asphalt shingles — in descending order of cost of materials and installation. Slate is beautiful. Properly installed with noncorroding nails over a

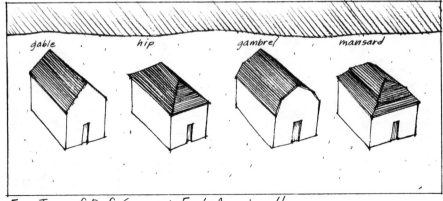

Four Types of Roofs Common in Early American Houses

frame strong enough to bear its considerable weight, it lasts practically forever. Slate roofs don't usually wear out. They usually come apart because the
nails disintegrate and the slates slide out of place. What turned me away from
slate for our house was its high investment cost in materials and labor-
intensive installation.

When the choice narrowed to wood versus asphalt shingles, Laura and I
had no difficulty picking wood for its texture, its look. I bought the shingles
from a distributor in Rhode Island. For the house, I selected number one,
taper-sawn Western red cedar, each shingle 100 percent heartwood, clear of
knots and vertical-grained. I also selected, for its appearance and durability,
a shingle with a butt end five-eighths of an inch thick, almost twice as thick as
most wood shingles. The difference between number one and number two
shingles is in the grain of the wood. A certain percentage of number two's
have a flat grain; some of them curl over time, exposing the roof to possible
invasion of moisture where the curl occurs. The difference in cost between
the two grades may be a wash over the lifetime of the roof, but I decided it
was better to pay a couple of extra thousand dollars for the best-quality shingles.

A rich reddish brown when applied, the red cedar shingles will weather
over time to a soft silvery gray with the kind of depth and subtlety of color
that one simply can't get in an asphalt shingle. Wood shingles are shipped in
bundles. The number of bundles needed to yield a "square" — a square is
the amount of material needed to cover a hundred square feet — depends

on the amount of shingle left exposed to the weather. It didn't take me long to calculate that I would need about fifty squares (or 250 bundles with the five-inch exposure I chose) for the three separate roof areas of the house.

Shingles need an underlayment to nail them to. The framers had already installed "skipsheathing" for that purpose. Many builders, as I've mentioned earlier, beginning in the 1960s and 1970s, began to install solid plywood decking over the rafters, then perhaps a layer of fifteen-pound felt over the plywood, and then the shingles were nailed through the felt into the plywood. That's fine for asphalt shingles, which aren't susceptible to rot from moisture. But it's not fine for wood shingles, which need to "breathe." Laid over plywood, wood shingles that should have lasted fifty or sixty years were rotting out in five or six. When this premature rotting began to occur, architects and engineers who noticed it first blamed the shingles. But it wasn't the shingles' fault; it was a side effect of the type of installation.

The way a wood shingle works is that as it absorbs moisture from outside it gradually swells until it becomes so dense that it is impervious to any additional moisture. Because the shingles expand in every possible direction, they are installed with a quarter-inch space between each one to permit sideways expansion. When the air outside the roof dries again, the shingle begins to dry and contract again. But the shingle may well be damp through to its underside before the drying begins, and if air can't circulate on the underside, moisture will remain there and cause the shingle to rot out from the underside. That was the problem I had feared with the roof of the house Laura had wanted to buy; it was constructed of wood shingles laid directly on plywood.

There is a relatively new product that the shingle distributor in Rhode Island discussed with me. It's a random matrix of synthetic fibers, three-eighths of an inch thick, which comes in rolls. Installed between shingles and solid decking, it compresses somewhat when shingles are placed over it and it is nailed to the decking under it, but it still creates a space for airflow under the shingles. The shingle distributor thought it would be overkill to install this material over my skipsheathing. I was glad to concur, since at a cost of about 50¢ per square foot it would have cost another $2,500 to install this additional underlayment.

You can find some very old houses today with shingle roofs still in adequate condition because the shingles were laid over planks instead of plywood and there were enough gaps and cracks between the boards to expose the shingles to some drying air on their undersides. Our skipsheathing was a return to this older and better underlayment. We would provide systematically for the gaps and cracks plywood doesn't allow.

Over the rafters the framing crew installed one-by-four ledger board — rough, not planed, to give us a full inch of board thickness to nail the shingles into. My son Bobby made their lives easier by carrying the material for skipsheathing from a pile thirty feet beyond the ell up to the main house attic — for which I'm sure he was remembered in their prayers. It was late October of 1992 when the framers began their work on the roof. They left an inch of space between each strip of ledger board. Every shingle would thus be exposed to air on its underside. Woody Woodward, using a measured block as a pattern, marked one side of a section of roof. Bill Delaney then nailed one end of each board in place as Woody rechecked the other end for straightness and nailed it. It took them half a day to strap one section of the main roof with ledger board — three times as long as it would have taken them to nail plywood decking in place. The structure looked less substantial and protective than a solid plywood decking would have looked, but the effect was to promise a longer-lived roof. So long as the shingles were properly overlapped above the skipsheathing, no water would drip through.

Jock Gifford came by while the framers were installing the skipsheathing and he watched their work with respect. "I can sense a plumbness, a regularity in all the work here," he said. "I could show you a lot of houses under construction where they wouldn't bother to measure the skipsheathing carefully. Something 'close' would be considered good enough."

Jock and I had earlier discussed several details of the design of the roof. Ordinarily you might look at a roof and think: What is there to design? One aspect is the overhang at the gable ends and at the eaves. There's a functional reason for the overhang: to keep water from dripping onto the siding at the gable ends or washing off the roof at the eaves and down the front or back walls. If you check this detail on many houses of colonial design, you'll find some in which there is virtually no protective overhang at the gables.

The spaces between the skipsheathing John Conrad is nailing over the rafters of the garage/office wing will let air reach the underside of the shingles to be applied over this underlayment, to dry the shingles after exposure to moisture and prolong their life.

Design for Box Return

The overhang also affects the appearance of the house. Without it, a gable end looks as naked as an untrimmed window. Jock's first sketches showed an overhang of three to four inches along the gable ends, six inches along the eaves of the main house and office wing, and ten to twelve inches along the eaves of the timber-frame ell. These dimensions had the right look and were sufficient to protect the siding from excess water.

Jock and I had also had discussions about the "box returns," the place where the bottom edge of the gable roof meets the eave of the front or back wall. In a casually designed colonial house there is nothing special about the trim at this juncture. But if the trim at the eaves turns the corner and extends a foot or more along the gable wall just beneath the "rake" (the board or molding along the sloping edge of the gable), the house looks better. Jock gave me some sketches showing several possible styles for the box returns. I chose the one I thought would look best with the gutter design, the depth of the overhang, and the length of the return. Each return had its own miniature hipped roof. The bead detail matched that on the rake of the main house; the bevel detail matched that of the rake on the office wing. I fabricated and installed the box returns and completed the eave and rake trim on the south gable of the main house to give John Conrad an example of the details and standard of workmanship I wanted for all the roof trim. I installed the box returns and trim on the other gables, and John completed the trim.

Even with the skipsheathing and trim in place on the main house and back wing, we weren't ready to shingle there. The metalwork for the gutters had to be installed. Mike Mullane, whom I had known for fifteen years, agreed to do the metalwork for me. Mike was a roofing contractor when I first got to know him, but he had gotten more specialized over the years. He developed a business selling snow guards for roofs, and he did custom copper work.

Some home builders elect not to install gutters, preferring to let water from rainfall or melting snow run over the eaves, counting on the overhangs to keep water from washing down the walls and to minimize the splashback against the foundation and lower walls of the house. But I believe that the value of preventing damage from splashback justifies the inconvenience of gutters — that they have to be cleaned of leaves and debris once or twice a year. I wanted to have at least one downspout or leader for every thirty-five feet of gutter, and to hook the downspouts into underground drainpipes to carry excess water away from the foundation.

Jock had drawn simple U-shaped gutters on the house plans. Mike added a couple of bends that wrap around a piece of bar stock at the upper front edge of the gutter. His revision of the design made the gutters stronger and closer to the shape of classic old wood gutters. Mike's design included a twisted bracket that was attached to the bar stock at the front of the gutter with a brass bolt and to the roof with brass screws. These brackets were to be installed every two feet along the length of each gutter to secure it at the eave. The drip edge is a metal panel that covers the first several inches of roof underlayment along the eaves. It sits under the first row of shingles and gives added weather protection to the edge of the roof. With the sample gutter Mike showed me, we'd have to install this separately. I asked him if he could make the drip edge an integral part of the gutter; he said that he could and that doing so would make a better system.

The gutters were made of twenty-ounce copper. The valleys, ridge caps,

The continuation of the cornice on the facade of this nicely weathered colonial house around the corner and a foot or two along the base of the gable wall is called a "return." It gives the corner a finished look. Our box returns, as sketched here, are slightly more elaborate but with the same purpose. We added gutters, which the photographed house lacks.

and other metal flashing to prevent water penetration were made of lead-coated copper. That meant that all of the gutter joints would have to be soldered. If you install vinyl or aluminum gutters, you can join them with rivets or screws and sealant. But I prefer the look of copper. I took the time to look at ready-made copper gutters, and I found one that would have been cheaper than having Mike make them for us. The ready-mades, however, came only in sixteen-ounce copper rather than the twenty-ounce copper Mike recommended. I thought about winters in our area. Snow and ice can impose hundreds of pounds of weight on gutters. Even with an extra bend for strength, I doubted that the lighter-weight copper would have the durability I expected for the price. I authorized Mike to fabricate the metalwork for the roof.

The roof metalwork, except the gutters, the valleys, and the ridge caps, which had to be soldered, was eventually installed by the roofer who installed the shingles. The shingle distributor in Rhode Island had recommended a firm from his state to do our roofing. I thought the commuting distance was excessive, but I didn't have a well-established relationship with any local roofers. So I did what anyone might do in the same situation: I consulted the local yellow pages.

Jim Normandin was having supper one night when I telephoned him; as you might guess, with my schedule I had to do most of my telephoning in the evening after I got home. Jim's wife answered the phone and told him there was a guy named Norm Abram wanting to talk to him. He thought she was kidding him, but as soon as he took the receiver he recognized my voice. I arranged to meet him the following Sunday morning at the construction site.

I don't remember now what held me up the day I was supposed to meet him. After years of dealing as a general contractor with people who failed to show or were late for appointments, I do my best to keep appointments promptly. I know how valuable time is to a tradesman. But for some reason I was almost forty-five minutes late, and Jim was just about to leave when I pulled up in my Bronco. He said he had a friend who is a big practical joker, and he had decided for sure that his friend had set him up again.

For three different reasons, I was almost immediately sure that Jim was the roofer I was looking for. His truck was immaculate, with a handsomely designed and painted sign on the cab door advertising his business. It was a stark contrast to the tar-covered vehicles common in the roofing business.

Jim's business card, secondly, identified the proprietors as steeple-jacks and listed an impressive range of services: roofing, gold-leafing, flag-poles, waterproofing, pointing, steeples, towers, and spray-painting. The notice in the yellow pages indicated that the firm was experienced in preservation or renovation work on old buildings; the album of projects Jim brought with him displayed work on many historical buildings in Lowell and Lawrence, Massachusetts, and other New England sites. I learned how much satisfaction he takes in working on old buildings. "Anyone can put up a step-ladder and paint a ranch house," Jim said to me one day, "but to take something apart and put it back the way it was a hundred and fifty years ago — *that's* rewarding!" Most of his projects involved work with slate or wood shingles. He was almost too talented to waste on a simple asphalt-shingle installation.

Finally, Jim looked ideally suited to being a roofer, one of the more hazardous aspects of construction. He was of medium height, very trim, and he moved with the ease and grace of someone who could clamber over roofs confidently. Jim gave me a quote on the roofing, and in less than a week I gave him the job. He brought a crew of five, more than he usually commits to a single job, but he knew there was a lot of surface to cover shingle by shingle, and it was already early November. By sometime in mid-December Jim wanted to be heading for Florida, where he has a second home from December to April. A couple of his men stay on through the winter dealing with service calls and emergencies. The remainder of the dozen or so men he calls on as needed collect unemployment insurance during the winter. "They don't seem to mind," he says. That, in a quick sketch, is the sociology of roofing in New England.

Roofers begin at the bottom of each section of roof and work upward. The first row — or "course," as the rows are called — is nailed in place with the exposed end extending slightly out past the drip edge, slightly out over the gutter. The second course is laid directly on top of the first one, but with the seams staggered so that moisture working down through a seam in the second course will hit a solid portion of the shingle below, not another seam.

Then, always working left to right, roofers work gradually up a section of roof, completing each course before they move on to the next. Five inches of each eighteen-inch-long shingle is left exposed; the remainder is part of the base for the following course. Each course has to be truly straight and parallel to the eave line or the installation will gradually go awry.

One way to establish the "line" for each course is to measure for it and then snap a chalk line. Jim had the kind of inventive little procedure that I admire in the best tradesmen. He tacks cleats to the roof sheathing above the area he is shingling and suspends a sixteen-foot piece of five-inch-wide pine as a guide for each course. He sets the bottom edge of the board on top of and flush to the butt edge of the previously applied course. This sets the correct exposure. Then he uses the top edge of the board as a ledge to prevent the shingles from sliding off the roof as he lays the shingles for the next course, being sure to overlap the joints and space the shingles properly before nailing them into place.

He also tacks a piece of strapping along the rake boards at the gable ends of each section of roof. The edges of the end shingles are nailed so they are flush with the outer face of the strapping; when the strapping is removed, the shingle edge is perfectly straight and with the correct amount of overhang.

On both sides of the main house roof, Jim and his crew installed some of Mike Mullane's snow guards — cast pieces of bronze shaped like arrowheads, riveted to straps of lead-coated copper. They had to be installed as the roof was shingled because the straps are set under the shingle courses in a symmetrical pattern Mike and I settled on together. The cast pieces are perpendicular to the roof surface and stick up about an inch and a half. They hold accumulations of snow and ice in place while they melt slowly, or they cause large sheets of snow and ice to break into smaller pieces should they start to slide off the roof, thus lessening the hazard to anyone standing below. Snow guards are very familiar devices on slate roofs, but in my opinion valuable also on wood-shingle roofs.

The skipsheathing on each section of roof was planned so that at the peak there was a continuous open slot about two inches wide. The tips of the highest few courses of shingles were trimmed so that the slot was not covered over. The roofers attached a ventilated ridge cap over the slot. Made of plastic, it has slots on both sides to let air exhaust from the attic; the slots are

baffled in such a way as to keep driving rain or snow from entering the attic. Our colonial ancestors didn't need this ventilation system along the ridge because they didn't have either insulation or tight enough construction to trap moisture in the attic, causing rot, mildew, and mold problems.

The fascia board behind the gutter also has a slot, about two inches wide, covered with screen to keep bugs out. Air enters the attic or the space between the insulation and the roof shingles at the lowest point of the roof and flows up and out through the ridge cap, removing moisture and heat.

There was only one element in this ventilating system that was difficult to design so as not to detract from the appearance of the roof. How should we cover the plastic ridge cap? The cap had to be covered because the plastic would deteriorate if exposed to constant ultraviolet radiation. Besides, I didn't want a plastic ridge cap atop a wood-shingle roof.

Ridge lines of old colonial houses were often covered with metal. I worked with Jim Normandin and Mike Mullane to design a cap. Jim installed the

The roofers should have gotten medals for working in such harsh winter conditions. At the end of each day, a protective tarp was placed over the unshingled section to protect it against overnight snowfall.

Mike Mullane (foreground) and his assistant attach a decorative lead-coated copper cover over the plastic ventilation strip at the ridge of the roof.

plastic vent, and Mike fabricated and installed the lead-coated copper cap over it.

The roofers started at the front of the house and worked toward the back. When they finished the main house, they moved temporarily to the back wing because the ell wasn't ready to shingle. The underlayment of the roof of the ell was more complex than the underlayment for the other roof areas. After a company subcontracted by Tedd Benson had made and installed the stress skin panels over the timber frame, we covered the panels, at Tedd's suggestion, with a layer of thirty-pound felt paper. Next we installed one-by-four-inch strapping every sixteen inches on center, running from the eaves up to the ridge of the ell roof. Over it we laid the same skipsheathing we had put on the other roof areas — running side to side.

There was now a continuous one-inch air space from the eave line to the ridge, providing a ventilating space between the stress skin panels and the shingles. This is technically known as a "cold roof" system and is a critical part of any roof installed over stress skin panels. The extra space was a blessing in disguise. It gave us room to run the wiring for the low-voltage ceiling lights in the ell. Otherwise I don't know how we could have wired the ell ceiling, because there was no attic space in the ell where we could run wires and then drop them down. As soon as the underlayment of the ell roof was in place and the conduits for the electrical wiring installed, Jim Normandin and his crew moved to that area of roof. When the roof was on the ell, they returned to the roof of the office wing.

It was a cold late fall followed by a snowy early winter. I had some big tarps, which we put on unshingled parts of the roof. Frequently, the roofers began their day by sweeping or shoveling snow off a tarp and then removing the tarp to clear an unshingled section for the day. Many nights there was a threat of snow, so the roofers replaced the tarps as their last work of the day. Applying wood shingles is a time-consuming process; add cold weather, snow, and tarps, and the pace slows even more.

As was his custom, Jim took time off during the winter to spend time at his Florida home. But he kept a crew busy on my project. On one of his trips back to New England to bid on some jobs, he found his crew nearly finished except for shingling the curved area of roof over the eyebrow-like window of my office. I hadn't finished the framing around the window, so the shingling above it was put off a few weeks. I asked Jim if he had ever done shingling around an eyebrow window, and he said he had done it many times, including the careful cutting, fitting, and nailing of slate roofing around eyebrows.

When I was ready later in December to complete the shingling over the eyebrow, I got Jim to come back for half a day to help me. The curved roof area above the window wasn't skipsheathed like the rest of the roof. The underlayment was plywood molded to fit the curve of the roof. To provide ventilation over the plywood and under the shingles, we installed a layer of the matrix material, giving me a small test site for the employment of this material over plywood decking.

Jim said it was the most enjoyable day of the whole job because he could talk roofing and construction with me to his heart's content. The most unforgettable thing he told me about the roofing business these days is that for every dollar he pays his employees, he has to pay another 91¢ for insurance. I think I got out of the general contracting business just in time!

An Optical Illusion

IF YOU LOOK at a well-constructed brick fireplace and then go outside and note the brick chimney that breaks through the roofline seemingly directly above the fireplace, it's logical to infer that they are the visible parts of a unit of construction that is made entirely of brick and is ramrod straight from its base in the basement to the cap on the chimney above the roof. But chimneys have a way of changing materials where they disappear inside the partitions of a house, or as they rise in a basement or attic, where the exposed surface is of less concern than in the living room or above the roof. Two of the three chimneys in our house are not entirely straight from bottom to top.

Lenny Belliveau and his assistant, called a "tender," came to our house in September of 1992 to begin work on the chimneys and fireplaces. Lenny's silvery gray hair suggests a man older than his roughly fifty years of age, but his face looks younger — maybe because he is remarkably good-natured and enthusiastic about his work. He grew up on a farm in New Brunswick, Canada, and loved farm life even though there was no electrical power at the farm until he was eight years old. His three brothers left the farm to learn masonry, and Lenny followed their example when he was eighteen years old. But a couple of years later he did what they hadn't thought to do; he brought his trade south to the States. After three years of working for other companies, he formed his own business and did commercial work, mostly masonry on nursing homes for the Commonwealth of Massachusetts. He didn't like the pressure or the paperwork, and seven or eight years ago he went back to constructing residential chimneys, fireplaces, and stonework, which he had learned in his first years as a mason.

During his commercial phase, Lenny once did the masonry for a commercial building in Somerville, Massachusetts. The Silvas — Phil, the father, and his three sons, Tommy, Dick, and Johnny — arrived to do the carpentry. Phil, seeing that Lenny had a forklift, offered to provide coffee every morning in return for the occasional use of the forklift to move lumber. Every morning at nine-thirty Phil served as many doughnuts as his crew and Lenny could eat, and coffee. Friendships were struck. When Tommy Silva became a regular contractor and carpenter on several *This Old House* projects filmed in the Boston area, he recommended Lenny Belliveau when there was need for a mason. That's how I met Lenny.

Lenny doesn't remember exactly which day in September he arrived to begin constructing the three chimneys, but he remembers that he spent twenty-one workdays on them and completed everything but the stonework of the fireplace in the ell and the laying of the hearths (not needed until 1994) by October 20, five days before his wife went into the hospital for surgery. The framing of the main house, including the roof, was complete, giving Lenny the basic structure in which to construct the first two chimneys.

Starting at the concrete floor of the crawl space under the main house, which had been reinforced with an extra-thick concrete footer where the weight of the chimney was going to be absorbed, Lenny built a shaft of concrete block and mortar from the floor up to the level of the first-floor living room. Then he made some forms and poured a horizontal concrete slab reinforced with steel rods, which sat on the concrete shaft and cantilevered out under where the hearthstone of the living room fireplace would sit. The architectural plans told Lenny what the level of the finished floor would be, and the slab was positioned so that when the hearthstone was laid on top of it, the hearthstone would be flush with the finished floor.

While the slab base for the living room hearth hardened, Lenny built a second shaft from the crawl space up to the first-floor library. He repeated the pouring of a slab to support a hearthstone there. Stage by stage Lenny went back and forth from one shaft to the other.

The next five or six feet of each shaft contained the masonry for a Rumford fireplace. Jock Gifford had provided detailed drawings for each fireplace, first for the framers, so that the openings through the floors would be accurate, and then for the masonry, to assure that all of the rules for a Rumford were followed. He and John Murphy did a lot of research to make sure everything was exact. Laura and I went to a masonry supply house before Lenny started the shafts to select the face brick (a different one for each of the fireplaces in the house except the family room fireplace, which is faced with stone, and yet another brick that would show above the roofline). Lenny, who knew exactly how much of each he needed, then purchased his supplies, facing the living room fireplace with "Suffield," the library with "Plymouth," and the bedroom with "Cambridge."

Lenny's estimation of Count Rumford's concept was the same as mine. These shallow, tall-for-their-depth fireboxes with angled sides really throw off the heat, and they are handsome as well. The tapered throat of the uppermost part of the firebox closes in to an opening to the smoke chamber that is only four inches wide. Originally, these colonial fireplaces had no dampers to prevent heat loss when the fireplace was not in use, or downdrafts when the firebox was cold. The small opening minimizes heat loss, but I planned to order some dampers custom-made for our fireplaces.

Lenny agreed with Jock on the importance of getting the right dimensions in a Rumford fireplace. It is not a particularly difficult fireplace to build, he said, but it must be constructed so that the opening to the smoke chamber is directly over the center of the firebox and directly under the center of the smoke chamber; otherwise it will not function effectively. Rumford fireplaces would not work in a low ranch house; they need at least twelve feet of chimney to create the conditions that make the chimney "draw," with the hot, smoke-laden rising air overcoming the downdraft of the colder outside air that is being sucked down the chimney. Certainly our first-floor chimneys were going to have well over twelve feet of flue. But what about the fireplace in the master bedroom on the second floor of the main house? It, too, passed the test. There were a few feet of flue behind the bedroom wall, topped by

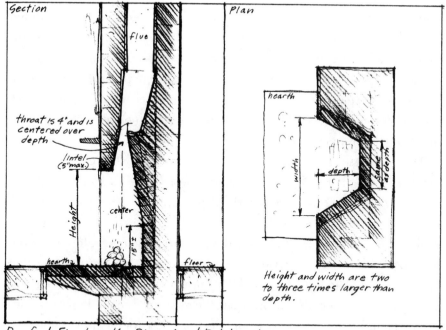

Section

flue

throat is 4' and is
centered over
depth

lintel
(5"max.)

Height

center

15"±

hearth

floor

Plan

hearth

width

depth

same
as depth

Height and width are two
to three times larger than
depth.

Rumford Fireplace Key Dimensional Relationships

ten feet of flue in the attic space, and then four more feet of chimney above
the roofline.

From the floor level of the hearth in the living room and library, the masonry structure was a combination of brick and cement block for the next
four or five feet. The floor and vertical surfaces of the firebox, and the masonry frame or face of the fireplaces between the firebox and the mantel —
everything that one could see from the rooms — was all brick, but there was
concrete block structural work behind the brick. Once what we call "the
fireplaces" were built, Lenny could remove the temporary supports in the
crawl space buttressing the hearthstone slabs, because the fireplace masonry
was a great weight on the other end of the slabs counterbalancing the way
they cantilevered out under the hearthstones.

As soon as Lenny had built the fireplaces up past the damper openings to
the smoke chambers, where the shafts are again hidden in the wall, he returned to cinderblock construction, which is structurally sound, less expensive in materials, and more quickly built — our colonial ancestors would
have used it, too, if they'd had it. For the shaft for the library fireplace, it was

a straight shoot from the first-floor library up between Lindsey's bedroom and the guest bedroom into the attic. But for the living room fireplace shaft, there was a complication. If the living room fireplace shaft rose straight, it would go right through the firebox of the fireplace in the master bedroom, not a desirable path. So once Lenny had the smoke chamber in place above the living room fireplace he had to begin angling the flue over so that it would bypass the bedroom firebox.

The flues inside the cinderblock shaft are made of baked clay. Each segment is two feet high. The segments are almost square in shape with gently rounded corners. Using an industrial diamond blade, Lenny cut flue segments so that he could angle the flue over into the desired path. No segment could diverge more than thirty-five degrees from the one before it, or the building inspector wouldn't pass it. When he got the living room shaft and flue up to the second floor, he repeated the construction of the living room fireplace for the master bedroom fireplace: the reinforced concrete slab cantilevered out under the hearthstone, and the brickwork for the firebox above it.

In all of this construction Lenny was very watchful about the air spaces between the various layers of construction. Although 90 percent of building inspectors in his experience would pass a shaft that left only one inch of space between masonry and any surrounding wood in the frame of the house, Lenny believes that two inches is the minimum to leave. Every inch of air space, he asserts, is worth four inches of masonry for absorbing and tempering the heat generated by fire in the firebox and by fire-heated gases in the flue. So there is air space between the clay flue and the surrounding shaft, and between the shaft and the wood frame of the house. There is air space between the layer of Glengarry dark paver brick lining the fireboxes of the fireplaces and the other blockwork of the shafts behind them. Each bit of air space is an effective agent in inhibiting the transfer of heat.

Once Lenny got the two chimney shafts of the main house past the second floor into the attic, he still had a lot of "corbeling" or reshaping and realignment of the shafts to do before they passed the roofline. One shaft had to be angled over about eleven inches so that it came through the roof centered on the ridge. Each shaft also had to be reshaped. Coming up from the basement, the shaft was five to six feet long and about thirty-two inches wide. For

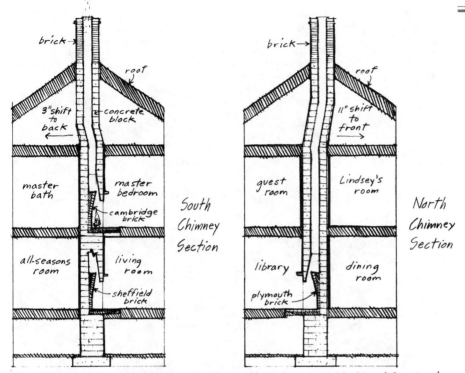

The two chimneys in the main house were built with cement block and four kinds of brick, and had to be angled to extrude through the roof at the desired locations.

the sake of being in scale with the house from outside, however, each chimney was to be four feet by thirty two inches above the roofline. In the attic Lenny gradually tapered the shafts down to the smaller dimensions specified in Jock's drawings.

Lenny did his first stone facing for a fireplace about twenty years ago. "A builder in Weston, Massachusetts, asked me to do one. I said I'd give it a try. Was I ever sorry! It took me over a week to do one fireplace. I lost my shirt on that job. Even now, I'll do a wall or a fireplace in stone and I can't believe how little I can do in one day. Even though I've been doing it for a long time, I'm still learning. You get better and better. With brickwork, you get better for a few years, but there's only so much you can learn. Stonework? There's no end to it."

The basic construction of the fireplace at the west end of the great room in the ell was similar to that of the fireplaces in the main house. A cinderblock

Lenny Belliveau has nearly finished the brickwork to line the firebox of the family room fireplace. The clay flue on the right side of the chimney will carry waste gases from the boiler in the utility room a floor below.

shaft rose from the basement floor to the floor level of the ell, where a con-crete slab was poured under the firebox and hearthstone. Visible brickwork backed by cinderblock took the fireplace from hearth level up to the begin-ning of the smoke chamber. Cinderblock construction resumed from the bottom of the flue up to the roofline, where construction reverted to brick-work for the exposed part of the chimney. This part of the fireplace installa-tion was completed in the fall of 1992 after Lenny had built the main house fireplace installations. The stone facing was held off a little until the stress skin panels had been installed on the roof of the ell as protection for any work inside the ell.

It was December when Lenny returned to install the stone facing. My first wish had been to use stone salvaged from the excavation to face the fireplace from the floor up to the sixteen-foot-high peak of the timber-frame ceiling. But Lenny judged none of that stone to be of satisfactory quality. So Laura and I went shopping. At a company in Sudbury, Massachusetts, we found a stone we liked named "Sterling," after the town of Sterling, Massachusetts, where it was quarried. While we were selecting the stone, the supplier asked who our mason was. I said, "Lenny Belliveau." "The best!" he quickly re-sponded, and said that Lenny had had a lot of experience with Sterling. Sup-pliers do know the best craftsmen.

You don't have to watch Lenny work very long at stonemasonry to realize that, as he says, brickwork and stonework are two different trades. Sterling is a very hard stone. Not a single stone in the entire facade fit the way Lenny wanted it to without some time-consuming shaping by hand using a mason's hammer and a carbide chisel. He knew very well the risk he took with every piece, that he might spend ten to fifteen minutes chiseling three faces of the piece to his satisfaction and then make a ruinous mistake on the fourth side.

The construction of what amounted to a vertical stone wall was like work-ing on a puzzle. As Lenny worked on shaping and placing a particular stone, he had to keep thinking about what would look good next to it and above it. He might, he says, look at a piece of stone on the floor ten times during a day without its ever seeming to be the right stone for the spot he is working on; and then, the next day, he will see that the stone he kept rejecting is perfect for a certain spot.

It took fourteen working days to complete the stonework on the great fireplace. I don't know whether Lenny broke up any of those days the way he sometimes does when working in the cold — laying some stones in the morning, taking off the afternoon to let the mortar dry and harden a little, and then returning in the evening to lay some more stones. Nor do I know whether nervousness ever led him to do what he admits having done on some jobs in the past, namely, returning after dark with a flashlight to check whether the joints were holding and beginning to dry and harden satisfactorily. On many of the days he worked on the stone fireplace, he and his tender were the only tradesmen on the site.

For the brickwork around the fireplaces and for the cinderblock in the shafts, Lenny used masonry cement. It has a fair amount of lime in it; the lime draws water from the cement mixture and allows it to dry and harden more rapidly. But the mixture of cement and lime means that the final mortar isn't as strong as mortar made entirely from Portland cement. Lenny's rule of thumb is that for joints that are thicker than half an inch he adds some Portland cement to the masonry cement to get added strength. For the fireplace wall he added one part of Portland to three parts of masonry cement. If he had used all Portland cement, especially in cold, damp weather, it would have dried so slowly that he wouldn't have been able to lay very many stones in any one day. He averages about six bags of cement a day, much of it mixed with a little portable mechanical mixer that he pulls to the job site behind his truck. He used a total of almost two hundred bags at our house, most of it mixed at a ratio of one part cement to four parts of sand.

I think of Lenny as being a pure tradesman of the old school. "Carpenters seem to have new tools invented every week," he says. "Except for diamond blades to cut faster and more cleanly, and for carbide chisels that make stone shaping faster, and for mechanical mixers to replace the old method of mixing all the mortar by hand, nothing in masonry has changed. You have to like it. It's still being done the way it was done fifty years ago."

When he does residential chimney and stone work, Lenny doesn't have to study the plans and then make calculations with a computer. He comes to the site, looks at the plans, thinks about it for five or ten minutes, and makes a bid. It isn't magic. It's experience and the basic elegance of his craft: a few

hand tools, relatively simple materials, and the acquired ability to do it both well and efficiently. He knows how much he charges per day for himself and his tender, he makes an educated guess about the cost of materials, and he has learned to project quickly and accurately how many days it will require to complete a specific job. That's all there is to it — several minutes of mental arithmetic and writing a bid figure on a piece of paper.

For the three chimneys and four fireplaces in our new house, Lenny gave me a bid of $15,000 inclusive of materials. When he had completed the job, he toted up his and his tender's hours and added the actual cost of materials purchased. It came to $14,870.

November 1992. The main house and the office/garage wing seem very detached from each other, but the timber frame for the ell between them will shortly draw them together.

Timber-r-r-r

IN ALL OF HOUSE building, nothing quite matches the drama of raising a timber frame. Our mental images of it often spring from films or photographs of barn raisings in rural societies. A community of Amish craftsmen gather and measure and cut the wood members using only hand tools. They fasten the beams and posts together with joints cut into the wood that are as varied and exotic as sailors' knots, and with wood pegs. They assemble the frame in sections on the ground; then every able body lends a hand to the ropes as the sections are raised into place by human strength and shrewd understanding of leverage.

The timber frame raising for the ell of Laura's and my new house differed in many ways from an Amish barn raising. The four major trusses arrived at the site having already been assembled at Tedd Benson's shop in New Hampshire with the employment of power tools, computer engineering, and high-tech glues. A hydraulic crane mounted on a truck bed lifted the trusses and other heavy members from the giant delivery truck to the deck of the ell as easily as if they were sections of a doll's house, enabling a rather small crew by Amish standards to raise our frame. Except for the trusses made in New Hampshire, our timber frame was built in place piece by piece rather than assembled on the deck in sections and then raised. There was no crowd pulling on ropes. Still, the frame for the ell rose in one day, and it had an elegant

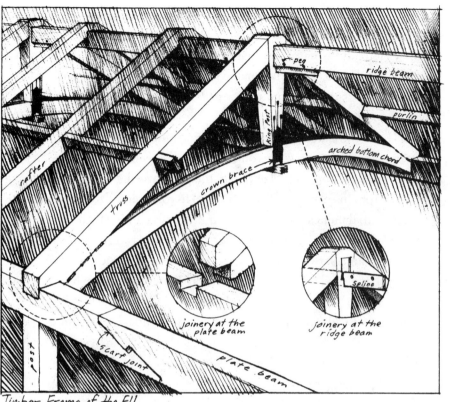

Timber Frame of the Ell

skeleton that no other part of the house frame came close to matching — so beautiful, of course, that the timber frame will be exposed for us to enjoy looking at every day we live in the house.

A few weeks before our day of timberframing in late November of 1992, I had been back in New Hampshire for a second visit to Tedd's shop. There were two purposes to the trip. The Benson Beam Team making our frame invited me to assist in the lamination of the arched bottom chord of one of the trusses, so that I would have the satisfaction of knowing I had had a hand in the truss construction. While in New Hampshire, I had to make a decision about the metal braces for the trusses. Tedd Benson had alerted me that if I didn't make that decision soon, the trusses couldn't be delivered for installation in November to complete the house frame before the worst of winter set in. It was just before Halloween when I arrived in Alstead, accompanied by photographer Richard Howard. I remember the date well because there was going to be a Halloween party for the Benson employees after hours. Somewhat to my dismay, I saw that someone had put a cover over one of the partially completed Abram laminated chords and planned to use it as a table to lay out the party food and drink. I hoped it would be a very quiet and decorous party!

Bill Holtz, the lead designer of the timber frame for our house, greeted us upon our arrival and showed us one of the arched chords that had been laminated the day before. Left clamped to the form overnight, it was now being removed from the clamps and fine-tuned. He volunteered to be our guide to visit a potential metalsmith for the truss hardware, after which we could return to Benson's and laminate another arched chord with the Beam Team.

It took us almost two hours to drive from remote Alstead to an even more remote country road in North Canaan, where Dimitri Gerakaris has his metalsmithing studio. His studio is in a tudor-style, timber-frame building on The Upper Gates Road, not a bad address for a man who has in fact fabricated a number of prominent gates. His work, I learned, is fairly evenly divided among residential projects like mine, commercial building projects, and public sculptures; his business card appropriately refers to his occupation as Sculptural and Architectural Metalsmithing. On a given day he might interrupt his work on panels showing the evolution of the locomotive for the Woodside Station of the Long Island Railroad to repair a local farmer's equipment. The one thing he doesn't do is shoe horses — not even the horse he himself owns and sometimes saddles and other times hitches to a carriage. "If no one in their right mind would be willing to undertake a metalwork project," he says, "I'm probably the right man for the job."

Dimitri is largely self-taught. That might be a prerequisite for his career, since there are so few craftsmen who practice his kind of art. His forge today is a spellbinding combination of the old and the new. He heats metal in his traditional coal-fired forge and shapes it by hand on an anvil with a 1916-vintage power hammer, but he also does high-tech welding and cuts steel with a plasma cutter — "real Buck Rogers stuff," he says. Yet for all of his credentials, I wasn't at first sure that this sturdily built man in a blacksmith's apron wearing also a full, well-trimmed beard (I'm partial to beards) was the man I was looking for to make hardware for our trusses. Tedd Benson had sent Dimitri a scale drawing of the trusses for our house so that he could think about the so-called stirrup braces that would hold the arched bottom chords and king posts of the trusses together, as well as the other metal straps the trusses would need.

The samples of his work that I saw in Dimitri's studio and in his portfolio were generally too ornate to look good in the simple colonial house Laura and I were building. Mentally, I was beginning to consider how I could back out of this meeting without hurting Dimitri's feelings. But he was thoroughly prepared for his meeting with me. I anticipated that he might have some rough sketches of possible brace designs to show me. No, indeed. He had actually fabricated part of a stirrup brace he thought would be handsome for our trusses. A "wrought-iron sketch," as he put it.

The sample brace was unveiled for my inspection in a deliberately ceremonial way. It had been engineered to be strong enough to fulfill its task of holding the arched chord up against the king post so that the arch wouldn't flatten under the stress it bore. But it was far from a simple clamp. The brace was to be fastened to each side of the king post so that the lower U-shaped section hung a few inches below the arched bottom chord. This space made the brace look very much like a stirrup at the bottom, justifying its name. Opposing cherry wedges would be driven together in the space between the bottom chord and the bottom of the brace to snug everything up; this wedge system was not an afterthought but rather a deliberately planned device to enable me to tap the wedges tighter whenever changes in weather altered the wood of the truss enough to need a little snugging. Dimitri had shaped the brace as it came up along the king post to look like a king's crown. The heads of the bolts fastening the brace to the king post would be worked to resemble jewels in the crown, every jewel hand-worked to be slightly different from the others. The crown had a very three-dimensional look to it. The brace was made of mild steel, the common steel used for automobile bodies, washing machines, and countless other everyday conveniences, a nice combination of sophisticated craftsmanship and common materials.

I was immediately struck by the imagination, the playfulness, Dimitri had brought to the design of a simple brace. The metaphor of crown to king post appealed to me. Everything looked fine but the size of the brace. It looked huge to me. "Isn't it larger than it needs to be?" I asked Dimitri. "You have to keep in mind," he replied, "that everyone will be looking up at this from floor level. There is visual foreshortening to take into account. It's the same factor that road painters allow for when they paint STOP on the road near an intersection. If you get out of your car and look directly down on the letters, they look exaggerated, stretched out. But in your car, as you approach from a distance, the letters look normal." To demonstrate his point, Dimitri took the sample brace and climbed up a ladder in his studio so that I could look up at the brace ten feet or so above me. He was right. It looked different up there, less imposing.

When Dimitri came down from the ladder, he told me he would have to work day and night to get the braces made when Benson needed them. Next he told me how much the braces and straps would cost. Then I knew for sure that I was dealing with a true artist rather than a simple country blacksmith! But I already knew that the great room of the ell with its kitchen, dining room, and family room open to one another under the trusses of the timber frame was our room of great indulgence. I placed the order.

It was getting close to the end of the working day before Bill Holtz, Richard Howard, and I got back to the Benson shop. There was time to laminate one of the arched bottom chords, however, because the process uses a mixed epoxy with a hardening agent that gives the woodworker only about thirty or forty minutes of working time before the epoxy sets. All of the strips of Sitka spruce for the chord were cut and ready for lamination. The spruce was not heated or steamed to encourage it to bend. In strips about an inch thick, it was naturally supple enough to bend against a curving form. We painted a generous coat of epoxy on both sides of each strip, using paint rollers to make the application go swiftly. The procedure was somewhat messy because as we clamped the strips together against the form, excess epoxy got squeezed out. The curvature of the form was tighter than the planned curve of the arch because the engineers at Benson's knew that there would be a certain amount of spring-back when the arch was removed from the form.

Bill Holtz mentioned to me as I rollered epoxy that the Sitka spruce is a good color match to the creamy Port Orford cedar from which the posts and

Once the Benson Beam Team and I had rollered epoxy on the strips of Sitka spruce to make a laminated, arched bottom chord for one of the trusses, we had about thirty to forty minutes before the epoxy set to get the strips assembled and clamped in the jig. It did require teamwork.

plate beams had been cut. Both woods stand up to moisture and sun exposure well, a factor worth valuing in a space with so much glass. The trusses and rafters were going to get two coats of a citrus-based oil before delivery. Over time the cedar and spruce will darken slightly to a honey brown, not quite as dark as pine.

I do a lot of gluing and clamping in my television workshop, but not often on the scale of the arch I was helping to assemble at Benson's. The time passed quickly and most enjoyably, and before long, having declined a warm invitation to stay for the company Halloween party, I was on the road for my fourth long drive of a single day. I felt I needed to get back to work to earn some money to pay for the day's indulgence.

Two days before Benson's was scheduled to deliver and install the timber frame, the weather took a nasty turn. Both snow and freezing rain fell by turn, covering the ground with a combination of ice and snow. It snowed again the night before the framers arrived. My son Bobby and I were the first to arrive that morning, and my first concern was that the trucks bearing the trusses and other parts of the timber frame might not be able to negotiate the driveway. We began to sand the driveway. John Conrad, the head framer for the

My son Bobby watches as John Conrad and I help hold and snap a line on top of the kitchen wall frame prior to installing a plate beam for the timber frame. The plate beams on each side of the ell were thirty-six

rest of the house, who was the next to arrive, pitched in to shovel more sand.

When the drive seemed passable, the three of us went over to the deck of the ell and erected some pipe scaffolding with a plank platform on top that could be moved along the deck depending on where the timber-framing crew was working at the moment. As I had done many times before in the foundation, deck, and framing stages, I decided to check the measurements again. Was the deck level and square? Was the already stick-framed north wall of the ell straight? Benson's crew was counting on finding actual conditions corresponding to the drawings on which they had based their cutting and assembling. Nothing, to my relief, was more than an eighth of an inch off; we could always compensate for an eighth of an inch.

Mid-morning was approaching, and none of the expected trucks had arrived. I was beginning to worry, but then the crane arrived, and I figured that once it was there, the others had to be on their way. The second Benson vehicle to arrive at the site was a pickup truck. Kurt Doolittle, the lead framer

for our project, and Kaspars Jaunzemis were its human cargo; in the back were the vertical posts for the south wall of the ell and the plate beams that would sit horizontally on top of the posts in the south wall and atop the stick-framing of the north wall. The posts were single pieces of six-by-six-inch cedar. But the plates were to run continuously more than thirty feet along each of the two walls — too long to be single pieces; they came in pieces that were cut to be joined by "scarf joints" — lap joints used to splice shorter pieces into one long timber. The scarf joints were centered over door openings in the south and north walls.

Since the trusses and rafters had to fit into mortises cut in the plate beams, the sequence of timberframing was to install the posts and plates first; the pickup truck had thus been deliberately dispatched earlier than the truss and rafters truck. Under Kurt's direction and with his and Kas's help, John Conrad, Bobby, and I set to work immediately on the installation of the north wall plate beam. That went quickly and easily. Then we turned to the posts along

feet long. Heavy timbers of such length are almost impossible to find, so the plate beams came in two pieces each, joined over the doorways with handsome scarf joints.

the south wall. Where each post was to sit, the deck was a layer of plywood. Under the plywood was a two-by-ten pressure-treated wood sill, which lay on the bond beam and the granite veneer at the top of the foundation. We cut away the plywood decking where each post was to stand, making a pocket in which the post could sit directly on the sill; the plywood sides of the pocket then prevented any unwanted movement of the bottom of the post. The masons had left metal straps embedded in the bond beam where the posts were meant to stand. Fastened to the outside of the posts, the straps will prevent the timber frame from ever lifting or shifting in a high wind or during the infrequent East Coast earthquake.

It took us about an hour and a half to place the posts and the plate beams. We had almost finished when, in a bit of happy timing, the trusses and other timbers arrived on a huge logging truck. The Benson contingent swelled to five — Kurt and Kas, who had been on the site for a while, Tedd Benson himself, Bill Holtz, and Rick Whitcomb. I admired the way Tedd treated his crew. He left them very much in charge of the installation. He was there as a resource in case any unforeseen problems surfaced, and as an extra hand.

The driver-operator moved the crane into position to airlift parts of the frame as needed from logging truck to house. Installation of the trusses and rafters began at the end of the ell next to the main house and proceeded westward toward the office wing. The setting of each truss had to be coordinated with the installation of a section of the ridge beam.

The ridge beam of our ell is intermittent, not continuous. The mortises in the king post of each truss and the sections of ridge beam are joined with a piece of cherry wood called a spline or free tenon. The spline passes through a mortise opening in the king post and then is joined on the far side to the next section of ridge beam. Oak pegs on both sides further strengthen the fit between the spline and the ridge beam sections. Structurally, the cherry splines and oak pegs are stronger than Sitka spruce against horizontal shear and other similar structural forces. Visually, the different color of the cherry spline where it is revealed against the Sitka spruce king post and ridge beam makes the joinery more of what Bill Holtz calls "a celebration."

I was as thrilled as any woodworker can be to watch the way the joints fit together as the timber frame evolved before our eyes. When the crane low-

The trusses, their metalwork wrapped in plastic, arrived on a logging truck. Lifted by a crane, they looked like giant feathered cranes as they glided toward their perches.

The whole timber-framing crew collaborate as a section of ridge beam is lowered into place on splines that connect the ridge beam to the vertical king post of the truss.

ered one of the trusses down on the plate beams, a tenon at each end of the truss would fit into a mortise on top of the plate so cleanly and snugly that we could almost hear the air being forced out of the mortise as the tenon went in. The huge timber frame had the refinement of fit that one would expect in a piece of fine furniture. Everyone watching expressed admiration as the trusses dropped into place and the rafters and "purlins" (horizontal braces between the rafters) were installed to complete the roof support system.

Daylight was already beginning to fade when the crew began placing the fourth and last truss near the fireplace. My parents and most of the other friends who had come to watch gave in to the bite of a very cold day and left for the warmth of their homes. As the crew worked on the final truss, Kurt Doolittle said, "Well, Norm, you've got to go get the tree. We need the traditional tree on the roof."

There's a custom in framing that once the frame is complete, the framers place a pine bough or a small tree at the ridge. The custom, called "topping off," has its roots in the construction of wood buildings, but sometimes I see the traditional tree perched on top of the steel girder framework of a contemporary high-rise building under construction. If I understand its symbolism correctly, the topping off ceremony pays homage to the gods of the woods and tribute to a safe and successful raising of the frame. When the fourth truss, and the ridge beam next to it, was fastened in place, the frame of our house would in fact be complete. Kurt was suggesting a perfect ceremony to cap an exciting day. The framers wished me to have the honor of nailing a small tree to the top of the ridge beam.

It wasn't difficult, nor did it take much time, to find and cut a small pine sapling nearby on our property. I carried the sapling to the ell and climbed up onto the pipe staging underneath the final truss. Richard Howard, our photographer, positioned himself on some scaffolding the mason had constructed by the fireplace; from there he had a good view of my ascent and could photograph the ceremony. Tedd Benson was standing about ten feet away on a plate beam, from where he, too, could see well. Laura and Lindsey and some of the framers gathered where they had clear views from the deck of the ell.

There was a stepladder on top of the staging, but its top step went only up to the bottom chord of the truss. I climbed up the ladder and stepped onto the truss. From there up to the ridge beam it was another five feet. As I stood on the truss, my head and shoulders were above the ridge beam but the rest of my body had to be hauled up.

Kurt Doolittle was already up on the ridge beam where I intended to nail the tree. I handed the sapling up to Kurt to free both of my hands to hoist myself up. Kurt took the tree in one hand, then leaned over to give me his other hand for an extra pull. With my left foot on the truss and my right foot on top of a cable strung just below the ridge beam to help the framers tighten the structure, I vaulted up toward the ridge beam. I knew in an instant that I was not going up —

I was going down!

To this day I remember exactly what it felt like. It seemed like the moment took forever. Actually, it was a matter of fractions of a second. I slipped out of

Laura, Lindsey, and I can't hide our excitement as the timber frame takes shape before our eyes.

Kurt's grip and fell over backward. It's just as well that I lost my grip on him because otherwise I might have pulled him down on top of me. I was sixteen feet above a hard wood deck, falling so that I would land head first. From that perspective, it looks like a long fall.

I believe I did what any falling cat would do: I tried to reposition myself in midair. I hit the staging on my left side, breaking my fall a little, bounced off it, and, in no time at all, was on the deck in front of the fireplace. Some who saw me fall thought I must have hit my head first. But I had one hand out far enough in front of me and had twisted enough to land on my shoulder and side first, sparing both my head and my legs the hardest impact.

My clothing cushioned the impact. I was wearing a thermal undershirt, a sweatshirt, and a down vest as defense against the cold, and my tool belt with only a few tools in it. The wool hat I was wearing offered some protection to my head. I didn't lose consciousness. I'm sure I was stunned, but I sat up, then jumped to my feet quickly to reassure everyone — and myself — that I was all right. All that I was immediately conscious of in the way of pain was stiffness in my left wrist from breaking the fall as I landed.

Richard Howard was so horrified by my fall that he didn't take a single photograph during or after it. One witness to the fall had an audio recorder on, but all it caught was the general pandemonium after the fall. Laura was very quiet and calm when she rushed over to me to assess the damage. Later she said one or two things about my intelligence in climbing to the ridge that I'm not going to share.

When everyone saw that I was not badly injured, people began to joke to relieve the tension. Someone suggested I restage the fall for the video cameras. Tedd Benson didn't joke about it. He knew how bad it could have been. All he said was "Good fall! Good fall! You did a nice job."

"In that precious second, that millisecond," Tedd said to me the next day, "I was sure that two people were going to tumble. Kurt is strong enough to hold you, but not when he's kneeling on a six-inch-wide ledge, holding a tree in one arm, and leaning over. I thought about everything at once. I was worried about your being injured. I was even worried about what Russ Morash would say when he learned that Benson's had let his star performer get injured."

It didn't appeal to me to try a second time to attach the pine sapling to the ridge. Kurt nailed it in place for me. The celebratory air of the day had been dispelled. Soon the framers drifted back to work in the semidarkness. Before they stopped for the day, they had set all of the timber frame but had still to tighten and tune it.

I was taping a television program the next day and couldn't be at the house when the Benson crew tightened the frame with come-along winches, set pegs, and checked the frame for squareness. I heard later that Kurt Doolittle took a tumble himself on the second day. Roger Hopkins hadn't finished setting the granite terrace outside the south-facing window wall of the ell. He had covered the ground where he still intended to set stone before winter with heavy, black plastic insulated blankets to prevent the ground from freezing underneath. From slipping slightly on one myself the day before, I knew that the plastic covering becomes very slippery when wet.

Kurt stepped on the plastic the second day carrying an armload of material, skidded, and broke a finger as he fell — at ground level. Except for a sore wrist, I seemed the next day to have suffered no painful physical consequences from my fall. The day after that, however, the second day after my

Kurt Doolittle turns a ridge beam into a gymnnast's balance beam. Give him a 9.5.

The timber frame was soon covered with stress skin panels, which insulate the ceiling of the ell and provide an underlayment for the roof.

fall, my knees were so sore when I woke up that I could hardly walk. I couldn't understand this stiffness. My legs, so far as I could tell, had borne the least of the impact. The only theoretical explanation of the fall I could conjecture was that I must have hooked my right foot on the come-along cable as I vaulted up toward the ridge beam.

Sometime during the second day of installation, the Benson team packed up and headed home to New Hampshire. The timber frame was complete. The frame of the entire house was complete. Now I was anxious to get the roof of the ell covered to protect the timber frame. The framers had wrapped all of the metal braces and straps in plastic so they wouldn't rust. But the sooner we got the stress skin panels on the roof of the ell, the sooner we could prevent the timber frame from becoming water-marked by rain or snow.

Tedd Benson had arranged both the fabrication of the stress skin panels and their installation — by a company and crew separate from his timber-framers. They came only a few days after the Benson crew had departed. Bad weather continued to plague us; it was raining the morning the stress skin panels were supposed to be installed. I expected the installation would be delayed and I couldn't be there that day in any case, but when I checked the new house on my way home after work, the stress skin panels were in place — rain or no rain. I was disappointed not to have seen any of the installation, which also used a crane to lift the panels from delivery truck to roof, but I was relieved to have the timber frame under cover.

I looked at the sky as I drove on home. Late November. Already the threat of a harsh winter. But I thought we could get the roof finished before the worst weather struck. John Conrad and Bill Delaney had promised to return as soon as the stress skin panels were in place to install a layer of thirty-pound felt paper on top of them. Over the paper they would install strapping running in one direction and skipsheathing at right angles to the strapping — all this before shingles could be installed.

I needed John and Bill now. I reached for the car phone.

Pins and Feathers

A CONTRACTOR — OR so I've heard — introduced Tom Wirth, our landscape architect, and Roger Hopkins, our exterior stonemason, to each other several years ago. The contractor is said to have remarked that the two men ought to get to know each other because both of them were "off the wall." I wonder how he came to link them temperamentally. It is in terms of talent that I link them. Both are artists. They have design affinities that make their approaches to landscaping an easy partnership.

Tom's manner is cheerfully upbeat. Roger is more of a loose cannon in mood and expression. When I think of Roger at work fabricating one of our walls or terraces, the phrase that comes to mind is "cutting and cursing." Or the other way around. Over time it becomes clear that Roger's sardonic asides are a thin veneer over an enthusiasm for stonework that is granite hard.

Early in our house-building venture Tom helped site the house, urging us to disturb the natural beauty of the forested site as little as possible. Roger subsequently did part of the installation of the granite veneer at the top of the foundation. In the latter half of 1992 their work became a collaboration as we began to think about a landscape design for the area near the house that Tom might design and Roger might install stonework for.

Roger came to stonework a little more indirectly than Tom came to landscape architecture. He received a degree in ornamental horticulture after he came back from Vietnam, and opened his own business emphasizing naturalistic landscapes. His main interest lay in the naturalistic use of water and stonework in gardens. He gradually came to specialize in landscape stonework. His elders were not convinced that garden design and installation was a fit occupation for a "Hopkins" until a relative doing genealogical research discovered an ancestor of theirs named Bartholomew Barba. In the early eighteenth century, when the rage in gardens was the formalism inspired by Versailles, Barba was a European garden designer extolling the virtues of naturalism. Once they heard the genealogical news, his family acknowledged that Roger must have gardens in his genes.

I told Roger about my enthusiasm for having stonework around any new house I might build even before Laura and I found a property we liked. I think he, more than anyone, knew how pleased I was that the property we bought had rock outcroppings, crisscrossing stone walls, an abandoned granite foundation, and chunks of granite lying about suggesting the quarrying of stone there in an earlier period.

During the summer of 1992, Laura and I began to talk with Tom Wirth about the basic topography of the landscape. We weren't ready yet to plan what trees and shrubs might grace the garden. What was on our minds was the layout of level areas, slopes, walls, and steps surrounding the house — particularly on the south and east sides, from which everyone approaches the house. Blasting and excavating the foundation had changed the natural contour of the slope on which the house sat more than Tom had wished. When Herb Brockert distributed some of the excavated soil and rock around the area nearest the house, the slope became much less steep. Yet the house was by no means sitting on a level plot.

From the beginning I had ideas and preferences of my own to factor into our discussions. For example, I strongly preferred to have stone terraces rather than wood decks adjacent to the house as social gathering places. A very high percentage of single-family homes of all architectural styles built today have one or more wood decks attached to them; an open wood deck, usually on the back of the house, is as popular these days as the covered front porch once was on houses built in the first half of the century. But open wood decks aren't typical features of colonial-style homes. The old covered

Pins and Feathers
An Old Technique for Hand-Splitting Stone

porches were known to the trades as painters' and carpenters' delights be-fore the development of pressure-treated wood, because they needed fre-quent repainting or restaining and, following rot or insect damage, had to be rebuilt every twenty or thirty years. Even with current use of pressure-treated wood and species of wood naturally resistant to rot and insects, I think of wood decks as a headache to avoid in the low-maintenance type of house we were building.

It was hard to imagine a garden at all when our talks began. The cleared area of the property was a crowded set of piles of rock, topsoil, wood from cleared trees, and construction materials. We knew, nevertheless, that even-tually the building materials would get used and the other piles moved out of the way or used or buried in rough grading.

On September 24, Tom Wirth submitted two plans for the garden. At first glance, the drawings looked very much the same to Laura and me. Both of them showed the space on the north side of the house — between the ell and the knoll — divided into a shade garden in the corner beneath the library and kitchen windows, and, next to it, a granite terrace just outside the family room. It is a quiet, sheltered place. I think our family and close friends will gravitate there often. It's a nice place to grill food and to watch and listen to the many varieties of birds who've already found the feeders Laura put out.

On the steep, rocky face of the knoll rising behind the terrace, there are ledges and niches that will accommodate a rock garden when we have time to plant it. There is a large flat area on top of the knoll, where the previous owner was going to build his house. Up there I intend to place a gazebo I have built.

It is on the east and the long south sides of the house that we found the conceptual differences in Tom's two plans. Because of the way the house is sited, one comes upon these two sides almost simultaneously when approaching the house. Together, they constitute the "front" of the house. The landscaping has to treat them in an interrelated way.

Both plans showed the same large granite terrace outside the south-facing window wall of the ell. The terrace extends out from the house to an imaginary line formed by extending the south gable wall of the office wing eastward. What this terrace does visually is unify the three distinct parts of the house — the main house, the ell, and the office wing. The granite terrace, which is adjacent to all three of the parts, says that this is one house, not three. Tom left planting-bed space between the terrace and the walls of the house; if she wishes, Laura can have an herb garden and flowers there just a few steps from the kitchen. Because the south gable wall of the main house does not project as far south as the south gable wall of the office wing, Tom extended the large terrace along the south wall of the all-seasons room in the main house just past its double French doors; this extension is not nearly as wide as the rest of the terrace. All four doors on the long south side of the house thus open onto a common terrace.

Between the terrace and the driveway Tom designed another garden area. It is almost two feet lower than the large terrace; the boundary between them is marked by a granite dry wall. There is another fairly steep three-foot drop in elevation from the south edge of this lower garden to the driveway. Plan #1 is what I call a plan inspired by history. It takes this lower garden and transforms it into a rectangular outdoor "room" imitating all of the other rectangular spaces in the house. The room is framed by a screen of trees and plants toward the front side of the house and by a line of boulders where the garden skirts the driveway. Four planting beds divided by granite stepping stones provide places for ornamental flowers and vegetables; the beds are like furniture arrangements in a room inside the house. It is an outdoor sitting room. This garden treatment is very much like what colonial homeowners might have developed in their front yards — a colonial garden for a colonial house.

Plan #2 seemed to us inspired more by the natural qualities of the property itself as Laura and I found it, and as Tom first saw it. All of the lines are less rigidly straight, a little more curvy and natural looking. Even the steps — from the terrace down to the garden, or from the terrace up into the house — have curved rather than sharply squared-off corners. Where the granite wall separating the terrace and the lower garden passes along the all-seasons room, Tom raised the height of the wall so that it becomes a sitting wall; he made the wall branch out at this point in a semicircular ending. This allows a larger terrace immediately outside the double French doors. I could happily imagine the pleasure of carrying a cup of morning coffee out to the terrace, sitting on the curving wall, and enjoying the sun rising in the east.

The lower garden in Plan #2 is more open and flowing than in the first plan. The stone walks meander. The landscape design says: this is a sloping, rocky site and we are honoring it the way we found it. Laura and I liked both plans, but we felt especially drawn to Plan #2. We liked its informality, which mirrors our way of life. We liked its emphasis on stonework. So we authorized Roger Hopkins to install the walls, walks, steps, and terrace as soon as his schedule permitted once Herb Brockert had done the rough grading. The pattern of the terrace stonework, as we see it, will present a strong visual pattern in the foreground for anyone in the great room looking out the glass wall; it will be a good visual counterbalance to the densely forested tract just beyond our property line. The terrace and the garden area beyond and below it allow a gradual transition visually in either direction between the driveway and the interior of the house. The floor of the ell is actually six feet higher than the driveway, but Tom's plan makes the difference in elevation almost unnoticeable by dividing the space between them into two levels.

One of the things I like about Roger's work is where he does it. As much as possible, some landscape masons cut and prepare their material off-site — usually in their storage yards — and then bring it to the site for installation and any final trimming necessary. Roger does his cutting where the job is, which gives his clients the pleasure of watching how he does it as much as

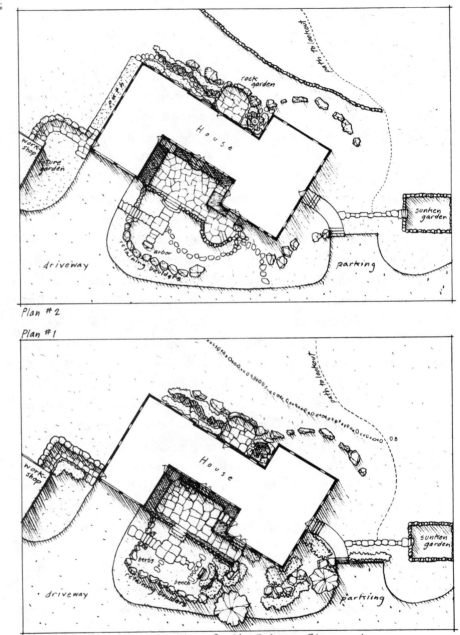

Plan #2

Plan #1

Tom Wirth's Two Alternative Plans for the Exterior Stonework

they'd like — or, in my case, as much as schedules permit. I could watch Roger cut and finish those huge stones endlessly. He began with the dry wall on the boundary between the large granite terrace and the garden below it. The large granite pieces he brought to the site were pulled from his three storage yards.

The large stones were cut into "one-man stones" or smaller. A one-man stone is what one man can carry and handle by himself. For the most part, that means a stone not heavier than a hundred pounds, although Roger says he can handle three-hundred-pound pieces as one-man stones — and I believe him. To cut the large granite stones into usable pieces for the scale of our wall, he used a technique dating from the late eighteenth century called "pin and feathering."

Along a line where he wants to split a stone, Roger drills holes several inches apart. In the eighteenth century all these holes had to be chiseled laboriously by hand. These days Roger can accelerate the pace by making the holes with a pneumatic drill. The holes have to be along one of the grains of the stone if the break is to be clean. Two metal shafts called "feathers" are dropped into each hole side by side. The upper end of each feather is bent over ninety degrees; the feather can't fall completely into the hole because the bend catches on the surface of the stone. A wedge called a "pin" is driven by hand between each set of feathers. The pin goes into the hole more easily between two metal shafts than it would if driven down against stone; metal against metal causes less friction than metal against stone. The stoneworker goes up and down the line of pins, tapping each one in turn to push the stone apart along the intended line until it breaks. The pin and feather holes leave hollowed-out marks called "strike holes" in the cut faces of the stone. Some masons prefer to hide them. Roger thinks they have their own appeal and likes to leave them exposed as a sort of craftsman's signature on the wall.

The tightly radiused, semicircular curve in the wall by the all-seasons room gave him fits. "I would have made that curve twice as big," he complained, "but I'm reluctantly following Tom's plan." He had to cut a lot of stones so that the side facing away from the terrace (on the outside of the curve) was wider than the side facing the house (on the inside of the curve). There was quite a bit of cursing during the cutting there. But from a design standpoint, Tom was right. The curve is just the right size to give an elegant finish to the line of the wall. Twice as large would have been half as interesting, designwise.

Roger often does his stonework alone. When he worked on the granite foundation veneer in August, he had his son, Jake, as his assistant. During the construction of the terraces and walls in November and December, he had two nephews, Michael and David Celwick, as his steady apprentices. Custom-fitting on the site is a great way for an apprentice to learn the craft. Roger agrees totally with what Lenny Belliveau, our interior mason, says about his craft — that nothing much changes. Many of the techniques are centuries old. The crux of the craft is to learn what master craftsmen have done before you. It's the practice, not the tools, that makes perfect.

For the flooring of the terraces, Roger recommended granite from a quarry in Westford, Massachusetts. The stone is tannish colored with a pinkish cast to it. The company had a problem with this granite because there was a "stain" running through it. The stain was caused by iron leaching through the stone over the course of several thousand years, giving the stone a rusty coloration, sometimes faint, sometimes pronounced, and irregular in its appearance and intensity. Most buyers consider the stain a flaw, so Roger, who thinks of the stain as an intrinsically interesting quality, bought the slabs at a hefty discount.

The terrace granite comes out of the quarry in slabs about nine feet long and four feet wide. Roger had most of the granite sliced two inches thick for the terrace floors; he also had some pieces sliced seven inches thick for steps. Despite this preliminary work at the quarry, Roger and his nephews were in for a tough installation. We all think of wood as having "grain." A sawyer can cut wood in any of several different ways (with the grain, cross-grain, quartersawn, etc.), depending on how he chooses to expose the grain visually. Granite also has "grain" — in three dimensions as compared with the two dimensions of wood grain. The grain in granite is not necessarily visible to the untrained eye. It has to do with the physical structure of the stone. If you want to cleave stone cleanly you have to abide by the grain.

The "head grain" is the hardest, so Roger had the quarry slice the granite for the terraces along the head grain. In New England the head grain runs

Forklifts are helpful in placing the largest stones in a dry wall. Roger Hopkins's nephew David guides the installation.

north–south in natural granite formations, almost as if it were magnetically lined up with the earth's poles. Roger could then make his cuts along the second grain, called "the grain," or along the third, the "rift grain," which is usually the easiest to split. The slabs come from the quarry marked with the direction of the rift. Unfortunately, in this granite, the rift was unusually hard. Instead of using just a chisel to cut the rift, Roger often had to start the cuts with his portable diamond saw. Even then, he sometimes got ragged edges. What he did was install the cleanly cut granite pieces on the front, or more public, terrace and the imperfectly split pieces on the back family terrace, where we might in fact prefer to have some small openings carpeted with spreading plants.

The granite floor of the terraces sits on a base of crushed stone. Roger was installing the terrace in late fall of a year when we got early cold weather and snow. He was worried that rain or melting snow might get into the gravel and freeze, so he covered the unpaved terrace areas with slippery plastic-covered thermal blankets. Water expands by 12 percent when it freezes, so frozen water in the gravel would have thrown off all their measurements. More than once Roger and his nephews had to clear snow off the blankets before they could move them out of the way to begin the day's installation of granite.

The granite was slippery, too, especially when wet, and something had to be done about that. The large slabs were sliced at the quarry with a diamond saw that leaves the sawn surfaces with a very smooth finish. If the slipperiness couldn't be remedied, I'd have to increase my homeowner's liability insurance considerably. Roger's nephews went over each exposed surface of terrace granite with a gas-fired wand, heating the surface of the stone to thirty-two hundred degrees Fahrenheit. At this temperature the less dense deposits of mica and feldspar in the stone melt and pop out of the stone, leaving tiny holes that break up the glassy smoothness of the surface, while the denser quartz, which is the predominant mineral of granite, is undisturbed. This firing process is called "spalling."

Tom Wirth's plan for the terraces suggested a very informal pattern to the stonework, so that anyone looking down on it would see the same kind of random pattern suggested by a century-old rural stone wall. Roger adapted that design to the kind of smooth-surfaced granite he was working with. His design as he installed it is slightly more "geometric" in feeling. The pieces of stone fit together so cleanly and closely that you have to stop and look to see that they are not joined by mortar or other grouting. Inside the great room of the timber-framed ell, I see what seems to me a virtual art gallery of wood and woodworking. Looking out, I see a landscape dominated by the beauty of stone and the wizardry of stonemasonry.

Before too long, the landscape will be softened by the plantings Tom recommends. He always begins his comprehensive design for a property by imagining someone approaching it for the first time. To create a sense of arrival, he has in mind a grouping of trees and shrubs on both sides of where the driveway begins. Between them the driveway quickly begins to curve in an S-shaped path through the woods toward the barely glimpsed house. Just as I conceived of the house as revealing itself only gradually, both outside and inside, Tom conceives of the landscape as revealing itself only gradually. A visitor begins with a voyage through woodland bounded by stone walls (which Tom wants to rebuild just slightly to undo some of the ravages of time

"Spalling" the south granite terrace with very hot flame roughens the surface just enough to make it less slippery underfoot when wet.

and neglect) and a series of intimate garden areas featuring ferns and other shade-loving small native trees and shrubs such as shadblow and viburnum.

The driveway brings a visitor in toward the front yard, the east facade of the main house, and a small parking area near the old sunken foundation. It is the most public part of the garden. Tom proposes that we fill in about half of the depth of the foundation, leaving a two-foot-deep garden with rough granite walls — a natural place for a hosta garden perhaps. He envisions a few simple plantings — broadleaf evergreens such as laurel and rhododendron — at the right, or north, end of the main house facade, to complement but not obscure the rock outcroppings on the knoll. At the south end of the facade, Tom proposes to extend the corner of the house with a group of trees and shrubs that will screen off the south garden beyond them as a more private area. It is this south garden with its terrace and lower garden for which he drew the two masonry plans, of which we chose the second. Some spreading evergreens such as juniper and yew will soften the edge between the driveway and the lower garden with its perennials and maybe even a few vegetables. A victory garden, perhaps, to invoke one of my producer's other television series. When Tom describes what the garden will be like in a few years — gardens grow even more slowly than houses — I see that it will enhance the house, that its maintenance will be manageable, and that it will complement rather than compete with its beautiful natural setting. Laura will be glad about the easy maintenance. Once I get my own shop at home it's going to be hard to get me away from my routers to mow the lawn.

A Mile of Tubing

**IF YOU INTERVIEWED** today all of the tradesmen who worked on our house and asked them to comment on their experiences, I think you would get a common reaction from several of them. They'd say it was a carefully planned house, that I knew exactly what I wanted, the price was fair, and they got paid on time. Everything was fine except the weather. The roofers would say that, of course, and the timberframers. But no one would say it with more conviction than the partners and their crew who installed the rough and finish plumbing, and the radiant heating system. As it turned out, they caught the worst weather of two different seasons!

The plumbing and heating firm I hired dates from 1986 when a master plumber and project foreman, Bob Blanchette, told his partner in their division of a mechanical contracting firm in Lawrence, Massachusetts, that he had decided to go into business for himself. His partner, Russell Des Roches, did sales and estimating work for the division. Both had degrees in civil engineering.

"I knew instantly that my work in sales was going to be much more difficult without Bob there to manage installations," Russell says. "So I said, 'Let's have lunch.'" Before lunch was over they had agreed to form a partnership. Blanchette and Des Roches was going to be too much of a mouthful, they figured, so they anglicized their gallic surnames into "White Rock." It was a big move for Russell to leave a large company and salaried income. "I didn't know where my next loaf of bread was coming from. You have to be disciplined in that situation."

Both were regular viewers of *This Old House,* which they watched especially to see what Richard Trethewey was showing in the way of plumbing, heating, and cooling innovations. Watching the shows got them interested in radiant under-floor heating. In 1991 White Rock got its first contract to install radiant heating. Thinking it prudent to have an advisor on their first installation, Russell and Bob went to Richard, who was an agent for tubing and controls for radiant heating.

"He's clever," Russell says of Richard. "He won't just sell you products and let you depend entirely on his talent. All the time he's talking to you, you're being interviewed. He checks your credentials and decides whether he wants to get involved with your firm. The New Hampshire job went slick. Then he saw another job we did, where he could judge our pricing and the quality of work we do."

In November of 1992, Richard, after suggesting White Rock to me as a potential plumbing contractor, loaned Russell a copy he had of the plans for our new house. Russell and Bob reviewed the plans and then came to look at the framed house with me. I sensed they were as good as Richard said they were, but I hadn't done any projects with them. So I suggested that they give me a firm bid for installing the rough plumbing for water and waste. Later I worked out a contract for them to install the heating system on a cost-plus basis. The materials for the heating system are relatively easy to estimate, but the labor is less easy to project. An estimator has to put plenty of cushion in a bid in such a situation to protect himself in case the installation requires more time than he anticipates. On the rough plumbing phase I saw that the White Rock crews worked very efficiently; on the heating system phase, therefore, I thought my interests would be served as well as theirs by a contract based on actual materials and labor costs plus the contractor's added percentage.

Whenever possible, a plumbing contractor likes to provide the fixtures as well as the pipe and fittings. His rationale has its points. The contractor has control of delivery schedules and can ensure that fixtures are on-site at the right time, well before installation usually, and that workmen are available to help unload the fixtures and move them under cover. Russell recalls situations where the homeowner has had something like a one-piece fiberglass tub unit delivered for a second- or third-floor bathroom after the stairways have been finished, leaving insufficient space to get the unit up the stairwell; homeowners, he says, don't like to be called and asked when they're going to send over a crane to hoist the tub to the proper floor.

Placing responsibility for buying fixtures with the plumbing contractor also clarifies those situations, in Russell's view, in which damaged fixtures are delivered or installed fixtures leak or don't function properly. The contractor becomes responsible for the problem-free installation of blemish-free fixtures. Nevertheless, I myself acquired all of the plumbing fixtures for our house. Laura and I consulted a catalog and then went shopping at a showroom, where we could see the line of fixtures made by the manufacturer I preferred on the basis of my past experience as a general contractor. White Rock was to furnish only the pipe and fittings.

The two things Russell and Bob needed from me were the floor plans and the manufacturer's specifications on the fixtures. These documents gave them information as to the exact placement of the fixtures and the kinds of connections to prepare between the fixtures and the water and waste systems. The three of us went over the house carefully, and I showed them what I wanted at certain critical places. There were some places where there was almost no tolerance for error; they had to put pipes exactly where the architectural draftsman and I had planned.

The utility room, where water comes into the house from the well, is in the basement under the pantry/laundry room — just off the garage. Copper pipes carrying hot and cold water to all the bathrooms and the kitchen, and water to and from the heating system of the main house, snake in one bundle from the utility room across the ceiling of the basement under the ell, held in place by hangers attached with screws to the floor joists above them. Some of the pipes destined for the second floor of the main house go up a plumbing chase tucked behind the curving wall where the staircase goes up from the downstairs hall to the second-floor hall. In the same tight space we had to put the plastic pipe bringing waste down from three second-floor bathrooms to the crawl space under the main house. A pipe to vent the plumbing system goes up there also and on through the roof.

In January of 1993, Bob and a helper began to measure and mark on floors and walls where the water and waste piping should go, beginning at the fixtures and working their way back to the utility room, all the while taking into account that they couldn't hide any pipes in the walls or ceiling of the timber-frame ell. They drilled and cut holes in subfloors and studs as necessary and attached hangers wherever pipes were going to be suspended from above. Their first installation was to bring the waste pipe down the plumbing chase from the second floor to the basement; from there it would go not back toward the utility room but forward, to the place on the front wall of the crawl space where it would go through the wall underground into the front yard and over to the septic tank. Since the septic system hadn't been installed yet, they brought the waste pipe within five or six feet of the wall opening and capped it off temporarily.

I think Bob and Russell were both amused and impressed that I had left a fairly long slot in the foundation wall, rather than a small hole of the approximate diameter needed, to allow the waste pipe to go through the wall. It was my experience, I explained, that no matter where a small hole was put, it would turn out to be the wrong place when the septic tank was set and the connecting pipe was ready to be hooked up. You can fool me once, I added, but I hope not twice. I thought I could see flickers of recognition in their eyes, signs of their remembering other jobs where the hole turned out to be in the wrong place and had to be redrilled through the foundation.

White Rock happened to be very busy when I wanted the rough plumbing installed, but Bob and Russell and their crew (usually three men, sometimes four, once five) were willing to work on our house on weekends and occasionally at night. There weren't any electrical outlets throughout the house to plug lights and power tools into, but there was power at the main panel in the utility room from which they ran long extension cords. It was a very cold January. The plumbers were all bundled up in heavy jackets, with extra sweaters underneath, hats, and gloves. In the spring I looked at Bob one day, and said to him, "Gee, you're a little guy. I thought you were a big guy." His

*Russell Des Roches
hosts an unforgettable
lunch for the plumbers
on a bitter cold day in
our unheated house.*

bulky January clothing had made him look at least forty pounds heavier than he is.

Bob remembers one day above all others from this period. It was a Saturday of one of the weekends the plumbers worked overtime. The temperature was below zero degrees Fahrenheit outside, and not much warmer inside the house. Russell brought along charcoal and a Dutch oven, and he cooked a meat pie with a pastry topping outdoors in late morning and made fresh coffee over the same charcoal. He also built a fire in the fireplace of the ell, pulled my picnic table close to it, and everyone had a hot lunch. Richard Howard came by to photograph their work and lucked into lunch. "Unforgettable meal," says Bob; their experience that day confirmed one theory, namely, that Rumford fireplaces are engineered to throw off a lot of heat.

The rough plumbing took about three weeks. Installation of the drainage pipe and fittings went rapidly. The lines for the water system took longer because all the fixtures are so distant from the utility room. Cold water is not a problem when the supply line is long, but it takes a long time for cooled standing water in an occasionally used hot-water line to clear and make truly hot water available to an impatient user. To remedy this situation, Bob suggested installing a recirculating loop for the hot-water supply. To make this loop required an additional pipe that would return water to the hot-water heater in the utility room for reheating as necessary. An in-line pump monitored by a temperature-control device keeps water in the hot-water line circulating to and from the hot-water storage tank so that hot water is always only a few seconds away from any faucet.

There wasn't anything unusual about the materials Bob and Russell used for the rough plumbing, except that the Massachusetts building code now prohibits plumbers from using lead-based solder on pipe connections — White Rock used a silver-based solder — and that they used a special type of cement that is workable down to zero degrees Fahrenheit to assemble the PVC waste pipe. Regular plumber's cement turns to something of the consistency of gelatin under very cold conditions, they've found. The oversize hangers they used to support water pipes enabled the insulation wrapped around the pipes to extend right through the hanger instead of stopping on one side of the hanger and resuming on the other side. My son Bobby, as stalwart as the plumbers in cold weather, installed the insulation around the pipes for me. Insulating cold-water pipes prevents them from sweating during hot and humid summer days; insulating hot-water pipes makes the water system more energy efficient.

The waste vents had been clustered in the house so that only two vents — one in the main house, another in the back wing — penetrated the roof. I planned the counter over the base cabinets on the east wall of the kitchen to be four or five inches deeper than usual; the cabinets themselves were standard depth. The difference between the depths of the base cabinets and the counter was just enough to permit tucking the pipes for waste disposal and for venting from the kitchen behind the cabinets. The main stack of plumbing pipes was conveniently located in the stairwell on the opposite side of the east kitchen wall. It was necessary to do the waste plumbing for the kitchen sink this way because we couldn't bring the vent through the stress skin roof without exposing the vent pipe in the kitchen.

Despite the harsh weather, the rough plumbing was accomplished with-out any — major or minor — hitches. Only one joint of the hundreds sol-dered wept and had to be resealed. The three bathtubs on the second floor of the main house were set during the rough plumbing phase and, as soon as we had running water, filled and tested against leakage — a test prudently done before they are closed in with any finish walls.

In the spring, Bob returned to do one job for me. I wanted to have running water. Since there was now no chance of a freeze bursting the pipes of an unheated house, Bob filled the entire water system — which up to this time had been filled with air — with water we could access from the outside sill "cocks" (the exterior water faucets threaded to provide a hose connection and located at the level of the sill at the top of the foundation).

In July of 1993, White Rock came back for a major piece of work. First, Bob installed the waste line from the newly installed septic tank back to the capped end in the basement. It was one operation he had to repeat. He in-stalled the line the first time with a forty-five-degree bend in it where it turned after coming through the foundation wall and headed over to the septic tank. Massachusetts code, Bob insists, permits that much of a bend, but the inspector wouldn't pass it. The inspector wanted a more gradual curve in the line. What the inspector wanted, the inspector got.

Then the White Rock crew began the installation of the heating system. When Laura and I first began to plan the house, I thought about installing radiant heating in all the floors, but I rejected the idea. Radiant heating would be ideal, I believed, wherever we had tile floors (bathrooms, the all-seasons room, the exercise room, the kitchen, and maybe the entire ell) but difficult to install where the floors were wood, and less effective under carpeting. Richard Trethewey and Jock Gifford talked me out of my reservations, with Jock being particularly helpful when he returned from a trip to Telluride, Colorado, and reported how radiant heating was installed under wood floor-ing in a house he saw there.

For the past thirty years, radiant heating has been a top-of-the-line system for residences and commercial buildings in this country. Hot liquid circulat-ing in piping under the finished flooring allows heat to radiate uniformly up through the flooring into living space. Russell says that with radiant heating a person can walk around barefoot in total comfort at a room temperature of sixty-eight degrees, but with other heating systems one would have to set the thermostat to seventy-two degrees to get the same level of barefoot comfort. The method works also in outdoor spaces, where the owner wishes to heat walkways or driveways enough to melt any snow or ice.

Radiant under-floor heating is more expensive to install on both the labor and materials sides than other hydronic systems — often two or three times more expensive. There should be some operating economies over other systems, but the initial cost differential is such that it takes a long time for radiant heating to "pay off" strictly on economic terms. Until the last decade, the worst knock on radiant heating was that the under-floor heating tubes, once made of copper or steel, deteriorated over time until they leaked. Tearing up finish flooring to replace them is inevitably a major expense and headache.

These days, the best tubing under the floors is made of a cross-linked polyethylene tubing that can withstand wide and repeated swings in temperature without becoming brittle or mushy. But White Rock and Richard Trethewey and others believe that tubing of dubious value is being used for some radiant heating and that when homeowners begin to report problems with radiant heating installed with inferior products, the method will fall under a shadow again. What one most wants to avoid in radiant-heating systems is a buildup of oxygen in the system. Oxygen diffusion barriers are built into the walls of the cross-linked polyethylene tubing to prevent oxygen from penetrating the tubing and rotting out the boiler unit quickly.

I haven't mentioned yet the second-most valuable aspect of radiant heating, next to the comfortable quality of warming it provides: its appearance, or, rather, the lack of it. No baseboard heating units to trip over or to catch dust. No free-standing radiators or other visible dispensers of heat. No living space cluttered up with heating equipment. Laura will appreciate the invisibility of radiant heating.

When I wanted the heating system installed in July of 1993, White Rock was already busy on another major job, installing the heating system of a new school. The work for the school had to be done before school opened in September. But I was feeling the same kind of pressure, wanting to get enough of the house finished that we could move in early in the 1993–94 school year. I wanted Lindsey to be able to begin the school year in our new town. White Rock set out to accomplish both jobs. For a while their crews worked on a double-shift schedule. They worked eight hours on the school, then came and worked the same length shift on our house.

July turned scorching hot just as the heating installation began. It was as hot as it had been numbingly cold when the crew did the rough plumbing in January. For several days in a row, the temperature climbed above ninety degrees. The workers kept a cooler filled with iced tea and soda on the site to prevent dehydration. "At the end of the day, I felt as though I had been beat up," says Bob, "but it wasn't quite as bad as a day I installed air conditioning in an attic. The temperature must have reached a hundred and thirty degrees. Some of the metal was so hot you could scarcely touch it with your bare hands." "It was work for a younger man," Russell says of our heating system installation.

The first step was to run one-inch copper main lines from the utility room out to the various manifolds or control panels. Through these lines a mixture of half antifreeze half water circulates through the heating system. From each control panel, cross-linked polyethylene tubing snakes out into its intended area of flooring and back. The maximum footage in each loop is about three hundred feet. Some rooms, therefore, required two or more loops. At the control panels there are adjustment knobs for the supply and return of each loop; one of the advantages of the type of radiant installation we were acquiring is that it permits a lot of fine-tuning by the homeowner room by room through different seasons and cycles of weather.

The system, designed by a distributor of heating system components and then modified by Richard Trethewey after I asked him to review it, divides the control panels by the type of finish flooring: wood, tile, or carpet. Each control panel services only one type of flooring. Some of the rooms have multiple heating zones in them; the great room of the ell, for example, has four different heating zones in one open space.

Radiant heating runs under the floor of the large shower stall in the master bathroom. It will be very pleasant on cold mornings to step into a shower and find the floor warm even before the water from the shower warms it. Above the double sink, or vanity, in the master bath, there is a large section of mirrored wall running from the vanity counter up to the ceiling. We ran a loop of radiant heating behind that mirrored wall, too. So long as the mirror is warm, it won't fog up even when a sink or the nearby shower or tub is throwing off steam. Russell reminded me that when people first began installing heat lamps in the ceilings of bathrooms, the purpose was first of all to heat the mirror over the sink so that it wouldn't fog over. But then people learned to enjoy getting out of shower or tub and toweling off under the lamp.

It took more than a mile of polyethylene tubing — somewhat between six thousand and seven thousand feet in all — for our house, laid out and clamped down to the plywood subflooring in very neat patterns. There is nothing haphazard about its placement. "You have to be fussy," Russell says about the process of installing radiant heat. If the crew didn't have any calluses on their knees before they began installing the tubing, they certainly had some after. Great care has to be taken that the tubing isn't punctured by any careless clamping. Operating our radiant system requires not less than fifty-five gallons of antifreeze and an equal amount of water circulating through the mile-plus of tubing.

There were more than a thousand joints to solder in the heating system installation. After the system was up and going, Bob found no more than a half-dozen joints that leaked a little and had to be redone: a very impressive performance.

In the utility room, White Rock installed a German-made boiler and hot-water storage tank coupled to the boiler. The tank has a coil in it through which hot water flows from the boiler to heat the fifty-five gallons of water available for sinks, tubs, and showers. Richard Trethewey first brought this heating equipment to my attention on one of our *This Old House* projects. Russell thinks it is ten to fifteen years ahead of competing equipment, although it costs about twice as much. The boiler comes with a twenty-year guarantee — many other boilers' guarantees extend only five to ten years — and the hot-water tank has a lifetime guarantee. The boiler is virtually self-cleaning.

There is a small computer in the control unit of the heating system. Most of the boiler and hot water equipment is wired before delivery to such an extent that the electrical hookup is simple. The crux of the system is that it

Copper pipes rise
from the basement to
the partition between
the master bedroom
and its walk-in closet,
where connections are
made in a control box
between the pipes and
the flexible
polyethylene tubing
through which the
heating solution
circulates in the
upstairs floors of the
main house.

The polyethylene
tubing is laid out in
carefully planned pat-
terns, then cleated
down with great care
to avoid puncturing it.

Yes, a mile of tubing. operates close to continually rather than firing up frequently and shutting off frequently, as many systems do. Continual operation enables the system to be effective in warming the house while running at lower than usual temperatures for the fluid circulating through the tubing. For every three degrees one can lower the average temperature of the heating medium, Russell calculates, the homeowner saves 1 percent on heating costs.

The coil inside the hot-water tank is made of stainless steel with a titanium coating, one of the features that justifies the lifetime guarantee. Water in an ordinary tank might lose one and a half BTUs per hour, but the tank we installed is so well insulated that it loses only a quarter of that.

A space-saving feature of the boiler and water tank installed in our utility room is that they stack. The boiler sits piggyback on top of the water tank. I'm sure Bob wished that the lighter water tank sat on top of the cast-iron boiler, which has a thirty-gallon capacity and weighs at least a thousand pounds. "I hope that water tank does last forever!" Bob said. He and one helper installed the stacked unit. "How the hell did you get it up there?" Richard Trethewey asked when he saw the completed installation.

What Bob did was attach "chain falls" to the floor joists above the unit and, with a come-along, hoist the boiler and swing it into place over the water tank. His ability to move heavy equipment regularly amazes Russell. "We had a cast-iron boiler that must have weighed three thousand pounds delivered to a commercial job. I went to do an errand, and when I came back, Bob and my son had the thing sitting in the cellar — down a bulkhead and around a corner. I said, 'How'd you get it there?' He said, 'Magic.' He won't tell me his secrets."

Except for someone occasionally showing up to install plumbing fixtures, White Rock had finished its work for me once the heating system was installed and tested. I had installed some radiant heating tubing myself under the garage floor before the concrete was poured; Bob connected the tubing to the system so that I can heat the garage in winter just enough to make it comfortable going to and from the cars, and to let winter ice slowly melt off vehicles while they stand in the garage. I didn't install any heating in the basement under the ell or in the adjacent crawl space under the main house. Since much of that space is below grade and the foundation is well insulated where it is above grade, I thought it would be fine to have the winter (and summer) basement temperature be about the fifty-five degrees that is the approximate earth temperature at the depth of our basements.

When James Sullivan, the plumbing inspector, came to make his final inspection of the plumbing, he found only one flaw. The soaking tub in the master bathroom has a hand-held shower attachment. Massachussets plumbing code says that wherever there is a shower, the water delivered to the fixture can never be hotter than 112 degrees Fahrenheit. Bob had set the general hot water temperature for the house fixtures at 135 degrees, which the dishwasher calls for. The same temperature water is delivered to kitchen faucets and laundry tubs and to bathroom faucets because users mix hot water with cold to their desired temperatures at those points.

The inspector had a digital thermometer with him. It looked like an instrument from a scientific lab. When he checked the tub with the shower attachment, he got a temperature reading of 130 degrees. "You can't have this," he said. "Where's the mixing valve?" "I didn't put one in that fixture," Bob admitted. Once again, what the inspector wanted, the inspector got. Bob installed a mixing valve behind the apron or trim between the tub and the floor to add the right amount of cold water to the hot water line before it gets to the faucet or shower attachment. It's reasonably accessible if you know where to look for it. It's one of the house's little secrets. Laura and I can now set the bath water temperature automatically to whatever warmth we want.

Let There Be Light

"I HAVE SOME CUSTOMERS I change light bulbs for. I have other customers I do quarter-of-a-million-dollar commercial installations for." So says Bob Russell, our electrical contractor. He's in his mid-thirties, just old enough to have learned the fickleness of the economy for a small businessman. Before New England's most recent recession, Bob ran three trucks and a crew of four electricians. At the moment, he's operating his business with one truck and one assistant.

After studying electrical technology at a trade school in Boston, Bob went to work for Joe Tremblay, whose joviality contrasts with Bob's terser, no-nonsense style. Later, Joe took Bob on as a partner in his electrical contracting business. Joe retired a few years ago but came out of retirement briefly to do part of the electrical wiring of our new house.

You might think, given my emphasis on power tools as a woodworker and carpenter, that planning and installing the electrical system in our house would be a major chapter in the story; in fact, it's not as prominent as some other aspects of the construction. What's involved in wiring a new house is fairly simple compared with the sophistication of many industrial and commercial installations — a hospital unit, for example, or stage lighting for a theater. Bob made an exception to his general practice in agreeing to do our electrical work; he usually doesn't wire new houses. "It's too boring," he says.

"New construction is usually done on a tight budget — you're bidding for pennies. There's no thrill in reading a set of blueprints and then doing a job as cheaply as possible. That's not our style of work. In the work we do, cost is usually not as much an issue as quality." Seventy-five percent of his business in the mid-1990s is in residential remodeling projects, where there usually are some novel elements to challenge the ingenuity of the electrician.

Several of the subcontractors I hired take the same approach to their trades as Bob takes to his. They try to limit their contracts to projects where quality is the first concern. In planning construction and hiring subcontractors, it is critical to have a sense of the quality you want in materials and workmanship and to be willing to pay reasonably for that level of quality. If you always accept the lowest bid without regard to the competence of the subcontractor and the quality of the materials, you'll sometimes get disappointing results, especially when the subcontractor runs into unexpected conditions and begins to look for ways to cut time and quality to protect his profit margin.

The first step of the electrical work was not done by Bob. It was done by me, with the welcome assistance of my son Bobby. When the foundation of the house was laid, and the deck and frame rose above it, some of the tradesmen used power tools even though we lacked electrical power drawn from a hookup to the utility power lines. I bought a portable electrical generator powered by a gasoline engine to provide temporary power. The machine was noisy, but it was effective.

One rainy August weekend in 1992, I began the process of bringing local utility power and services to the house. I rented one of the mini earth-moving machines with a backhoe attachment. Bobby and I dug a trench for the electrical, telephone, and cable television lines. The nearest place to tie in to underground utility lines was along the northwest property line, where a fiberglass pad with a transformer mounted on top straddled the property line between us and our nearest neighbor. There was access to telephone and cable lines next to the pad. Each homeowner has responsibility for bringing these services from the tie-in location to the house.

It was about 240 feet from the transformer pad to where I wanted the electrical service to enter the house — a long enough distance that when

Bob Russell later pulled the electrical wires through one of the conduits I buried in the trench, he had to hitch his truck to the wires to get enough pulling power. I brought the telephone and cable wires into the house at a different spot — on the north foundation wall of the garage — which required a run of only 220 feet.

Bobby and I and the mini-excavator got down to the twenty-four-inch-trench depth required by the electrical code, but it was tough digging. We scraped rock ledges most of the way. I still remember the discomfort of working in wet clothes to dig soil that turned to mud as we moved it.

We laid a few inches of sand in the bottom of the trench as a bed for four conduits — two two-and-a-half-inch-diameter conduits for electric service; and two one-and-a-half-inch-diameter conduits — one each for telephone and cable television service. Very likely I could power the shop I hoped to build behind the house off the house's electrical service, but just in case the power supply turned out to be inadequate for both buildings, I laid a second conduit for any additional future service. An extra conduit is a lot less expensive than renting a machine to retrench; besides, I wanted never to have to repeat the soggy trenching operation. Lightning may not strike the same place twice, but rain often does.

The PVC conduit comes in ten-foot sections. Bobby and I glued them together, installing a pull line as we assembled them for the convenience of anyone pulling wires through. To conclude the weekend mudfest, we covered the conduits with a few inches of protective sand. The final backfilling of the trench with the soil that had been excavated could not be done until the electrical inspector had approved the trench work.

Bob Russell advocated bringing 200 Amp service into the house, with the capability of later boosting the service to 400 AMP capacity if we desired. "Some people might argue for putting in a 400 Amp service at the beginning," Bob said to me in our first discussion of the electrical service. "But I think it would be a waste. Even when your woodworking shop is up and running close to the house, it will be mostly a one-man shop with one power tool in operation at a time. Even though you'll have four different air-conditioning units, they're not all going to be making equal power demands at the same time. The heating and hot-water systems won't tax a 200 Amp

Even though his work running power lines will be hidden behind blueboard, Joe Tremblay is meticulous as he lays out the wiring that will feed power to the upstairs hall and stairwell.

service. As large as the house is, there will only be three people living there. How much power can three people use at one time?" I found his argument persuasive, particularly since we could boost the service later without re-trenching.

After Bob got the electrical wires pulled through the conduit, it was sev-eral weeks before the local power company activated the connection to give us electrical power at the panel in the utility room. Most of the rough framing was accomplished using the generator as a power source, but it was a joyous day when John Conrad and his crew could just "plug in."

Except for a brief reappearance when we got a power hookup, Bob was not on-site from early September until early December, when I needed him to install conduits above the stress skin panels covering the ell for the wiring to power the ceiling lights in the ell. Bob had to get his conduits laid quickly in the air space because as winter approached I wanted Jim Normandin, the roofer, to get the shingles quickly installed over the strapping — a step that would shut off further access to the air space. Bob cleared snow off the plas-

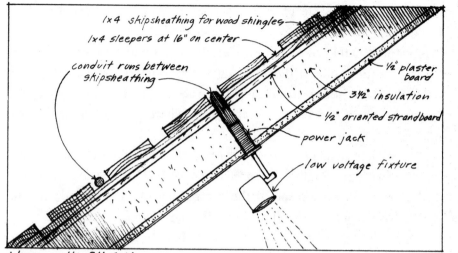

Wiring in the Ell Ceiling
We ran electrical wiring between the stress skin panels and the roof shingles to get to the ceiling fixtures in the Ell.

tic tarps covering the strapping a couple of mornings before he could lay more pipes to carry wiring for low-voltage ceiling lights, but he got the pipes installed and foamed in for protection against moisture. Then he disappeared until the plumbers had finished roughing in the waste and water systems.

"We like to come in after the plumbers rough out their system," Bob says, "at least after they have installed the large pipes for the waste system. They make big holes. Their pipes are straight. It's almost inevitable that if you put wiring in before rough plumbing, some of the wires will get cut — accidentally, of course! — or the plumbers will run into a wire that happens to be where their pipe *has* to go."

This sequence meant that the plumbers had to drag extension cords with them all the way from the utility room, and work in the coldest part of the year in an unheated house, while Bob Russell came back in somewhat better weather to complete the rough electrical installation. The first two places he wired were the basement and crawl spaces, and then the attic. Some electrical contractors, Bob says, work in a different sequence. They wire the basement and attic last. "But then the electrician spends a lot of time crawling around in the dark with flashlights, after which the insulators come in and work in the nice light the electrician has left behind. I wire the dark places first because I know I'm going to be in them regularly, dropping wiring down to other parts of the structure from the attic or pushing it up from the basement. Of course, one of the advantages of working on new construction rather than renovation is that the walls aren't covered yet. The frame is open, so you don't have to fish for wires the way you have to when wiring existing walls."

Except for installing lighting, much of the work of the electrician — once the rough wiring is installed — gets tied to the work of other subcontractors. As the heating system is installed, the electrician has to come in and perform the electrical component of the job. The radiant heating system we chose is operationally very simple. It is basically a self-controlled system that monitors outdoor temperatures, indoor temperatures, and the temperature of the water in the system. There is one main control, which can be adjusted to kick the basic indoor temperature up or down, and there are stations where the radiant heating for a set of rooms can be fine-tuned. But the system operates without electrically powered thermostats on interior walls. Just to be cautious, I did have Bob Russell put in wiring for a thermostat system, but Richard Trethewey, my heating consultant, and I are both confident that thermostats will never be installed in the house.

When we decided to install an air exchange system to guarantee that fresh air would be introduced by a timer at intervals into the house, Bob came back to do the wiring for it. The building code in Massachusetts requires that all new bathrooms have ventilation systems, so that when you turn on the light in the bathroom, the ventilation fan turns on. Bob wired the bathrooms so that the air exchanger provides bathroom ventilation and is activated by turning on the bathroom light. Fortunately, there wasn't much duplication of work required because of changes or additions as we built the house, but there were many times when Bob and I conferred about how to handle placement of an outlet or a switch.

The electrical plan that Jock Gifford prepared for our house recommended the location of lighting fixtures and their controls. Before Bob Russell did any of the rough electrical work, I brought in a lighting designer to consult with us on the lighting plan. The way it worked out was a happy accident. When Laura and I were first talking about building a new house, I mentioned our intentions in a talk I gave at a builders' association meeting in Massachusetts. Afterward, a young woman introduced herself to me, said she would be thrilled to have a chance to design lighting for the house, and gave me her card — which I tucked away. Over a year later, I recognized the woman, Melissa Guenet-Zagorites, when she was hired as lighting consultant on the renovation of a house we were videotaping for *This Old House.* I looked in the stack of professional cards I have acquired as potential construction resources and there hers was.

When I think about prevailing practices when my parents built their dream house compared with what I want for Laura's and my house a generation later, I see that many things have remained much the same. But lighting is an aspect of home design that has evolved rather remarkably. In my parents' day, a lighting plan involved wiring for central ceiling fixtures in most rooms, a little task lighting around the sink and counters in the kitchen, and plenty of electrical outlets in all of the rooms into which portable lamps and other appliances could be plugged.

In my view, the quality of light in and around a house at various times of day and night has tremendous impact on how much the house is enjoyed. Lighting determines the ease with which work is done. It affects mood. It influences personal safety and security. Inadequate lighting hastens fatigue. Yet in many homes lighting is not very sophisticated, undermining the capacity of the house to look its best and function well.

Melissa's standard for lighting a house is that it can be lighted comfortably for every purpose by installed rather than portable fixtures, most of them recessed into ceilings or otherwise unobtrusive visually, and that the lighting should have a little theatrical spark to it. I was tied up with other work when Melissa came to walk through the frame of our house with Jock Gifford and Laura in November of 1992. The lighting levels she recommended room by room were roughly the same as recommended in Jock's plan, but her initial ideas of how to accomplish that lighting with flair were convincing enough to Jock that before she left he asked her to design lighting for his house on Nantucket. For me, Jock's interest in having Melissa design lighting for his own house was comparable to having a *Good Housekeeping* Seal of Approval.

Soon after the walk-through, Melissa invited Laura and me to a demonstration of lighting in a model room at the electrical supply house where she worked. It was a revelation to see the cumulative effect of three kinds of lighting in the room — ambient, task, and accent. Ambient lighting provides a desired level of general illumination in the room. Task lighting is directed to places, such as a kitchen counter or a desktop, where close work is done regularly. Accent lighting makes one aware of special features in a room — for example, flower arrangements, artwork, or a special area rug. If all of these lighting systems are on when one walks into a room, one's eye doesn't distinguish one from the others very much. But if one turns them on one by one, the effect is astonishing. Ambient lighting makes a room pleasant and livable, but its comparative uniformity is less than dramatic. The addition of task and, especially, accent lighting adds brilliance to the lighting composition.

Melissa said she sometimes consults on the lighting of houses that have virtually been designed around the furniture. Laura and I were at the other end of the spectrum. Our house had to be wired for lighting, and fixtures selected and installed, before we made many decisions about furnishings and furniture arrangements. Laura and I were too busy with construction decisions and choosing materials such as floor coverings to spend much time discussing interior decoration. So Melissa had to make educated guesses in some rooms concerning all three kinds of lighting. She related ambient lighting to where furniture was apt to be placed, recommended accent lights for walls where we would probably hang artwork, added task lighting where we were most likely to need its enhanced power.

Some of her suggestions I would never have thought of myself. She made sure, for example, that a light was positioned to illuminate each hearth, so that the different colors of each piece of marble or granite would be discernible; otherwise the stonework all looks black after dark. She lighted the handrail of the stairs in the main house, angling the light so that the balusters will throw nice shadows for dramatic effect. She positioned three ceiling lights in the great room to make pools of light in a path extending from the main house hall at one end to the mud hall at the opposite end of the room; with dimmers, these lights become a softly but effectively lighted path for anyone crossing the room after the other lights have been turned off for the night.

I haven't dared to count all of the lighting fixtures that we eventually installed in the house. Bob Russell says there were about 280 of them in the plan at one point. But we kept tinkering with the plan. There are some lights that we wired for but may never install. There were lighting fixtures dropped from Melissa's plan. And there were fixtures added as Melissa, Bob Russell, and I thought more about the plan. We added nine more ceiling lights to the kitchen area in the ell, for example, four of them as task lighting above the work peninsula that divides the kitchen from the family room, and five of them to illuminate the medium-dark cherry wood of the kitchen cabinets so the cabinets reveal their natural wood hue at night rather than looking too dark.

In all, there are more than seventy light fixtures recessed into ceilings in the house. These are low-voltage lights with halogen bulbs, which require transformers to adapt the basic electrical service down to the capacity of the fixture. The rough-in portion for the recessed fixtures has a built-in transformer, so all the electrician has to do is run line voltage to the fixtures. The opening of each of these recessed fixtures is only two inches in diameter, but the fixture throws a pool of light as wide as three feet on a nearby surface. When the recessed fixtures are first installed and you look at them in an unfurnished room, they are obvious though not intrusive. When the room is furnished, however, the fixtures seem scarcely noticeable.

In a number of locations, including the ceilings of the ell and my office, Melissa prescribed focal jacks rather than recessed fixtures. The jacks have small heads mounted on stems of adjustable length. The focal jacks are really just sockets to plug the heads into. Since they do not contain transformers, we had to find convenient and central locations for remote transformers to power the heads. In the attic space above the bird room in the back wing of the house, there is quite an assembly of transformers! A few other transformers are located in the crawl space or attic of the main house. Each of these fixtures is small, but it throws an impressive pool of light. In the cathedral ceiling of the ell, these jacks were located to be as inconspicuous as possible so that the fixtures would not draw attention away from the timber frame.

Melissa left the placing of switching and dimmer controls for the lights to

Laura and me and Bob Russell. We had to think through how the various rooms were going to be used and, therefore, what the best placement was. Since the switches are not labeled, we tried not to make any bank of switches too intimidatingly complex. There are quite a few switches in some locations, such as the ell, so we deliberately chose a switch that is very shy and does its best to fade into the background. Sooner or later every room in the house will be subject to use by someone who is unfamiliar with the switching arrangement but wants to turn on certain lights and leave others off. In the kitchen area of the ell, there are several banks of switches controlling a number of interior or exterior lights where it really helps to know the arrangement; otherwise the switching controls are going to frustrate anyone using them.

The interior lighting was being installed before we turned our minds to outdoor lighting. Melissa gave us some suggestions for exterior lighting as well. The front door of the main house is flush with the facade rather than recessed, so there was no place above the front door to mount recessed lighting fixtures. We chose to install a "conductor's lamp" on each side of the door. The lamps are hand-produced locally in copper, which we will let age naturally. Though these lamps are not colonial in origin — they were invented in the nineteenth century, as train lines crisscrossed the country, to be carried by conductors and hung on posts at the edge of tracks when not in use — they look fine on colonial houses. Their old oil-burning wicks have given way to modern bulbs. A third, smaller conductor's lamp illuminates the entry from the main terrace into the mud hall in the office wing and three more illuminate the garage doors.

To augment the lighting of the main terrace outside the south wall of the ell, Bob Russell installed a row of recessed lights in the soffit (a "soffit" is the exposed undersurface of any overhead component of a building, such as an eave or cornice or arch) under the edge of the roof. These lights shine down on the flower beds just outside the window wall. He also installed five low-voltage accent lights on the trim of the south gable and southwest eave of the main house. Each light illuminates a different area of the terrace below. This lighting has three distinct functions: it has a security function; it enables a person inside the house to see the beautiful stonework outside and the terrace plantings; and it transforms the appearance of the window wall from inside after dark. By daylight, the window wall is a visual invitation to look outside at the natural setting of the house. But unless there is exterior lighting, all windows from inside at night are black holes, and a long window wall is a very large black hole.

On the north side of the house Bob installed a lighting system similar to the one on the south side. Soffit lights shine down on the granite terrace. Standard twin-head floodlights mounted high on the north eave of the main house and on the north eave of the office wing illuminate the knoll. In the front yard to the east of the house and in the garden between the terrace and driveway on the south side of the house, Bob buried electrical conduit in the ground, which, after the garden is further developed, we can utilize to extend exterior lighting farther from the house toward the surrounding woodland.

There always seemed to be something more to call Bob back to install or adjust. From the time he first worked at the site, pulling the wires to bring electric service from the transformer to the house, he returned again and again right up to the day of the inspection to get our occupancy permit. A couple of hours before the inspector arrived, Bob hurried about installing the last few lighting fixtures.

When we have lived in the house for several months, giving us the opportunity to become familiar with the lighting and to furnish our new home more completely, Melissa will return and tune the lighting system, refocusing some of the fixtures, changing the elements or bulbs in some of them to make them softer or more intense as we judge best.

My son Bobby installed the final phase of the house's wired systems — the telephone, cable television, and audio systems. His regular job while we were building the house was doing telephone and data hookups for a large local computer company. While the frame was still open, he and I marked locations in each room for telephone jacks and cable television connections. Wires for these systems ran from the rooms back to the utility room. The telephone and cable companies supplied the service wires and pulled them through the conduits underground to our utility room, and made the connections there; but it was up to us to extend the wiring through the house.

On each side of the front door we installed copper and glass lamps inspired by oil-burning lamps train conductors carried in the nineteenth century.

In older homes the telephone cable is usually run as one continuous line from room to room; it contains only four wires, which limits the system to about two telephone numbers. For our house Bobby ran a cable with eight wires from *each location* back to the terminal block in the utility room. A staggering number of combinations are now possible. Not only can we have a sophisticated telephone system; we can network our computers in various locations in the house to each other or to a central printer. The cable television system is similar to the telephone system in that a cable runs from each room back to the utility room, giving us a number of options for sending or receiving video signals throughout the house.

Wiring the systems this way uses a lot of wire, but, relatively speaking, wire is inexpensive. There is no better time to install it than before the wall studs are covered with insulation and blueboard. My only fear is that home electronics is developing so rapidly that it will outpace our planning and installation. Bobby bought all the supplies and did the installation on weekends. As Bob Russell had earlier done for the electrical wiring, Bobby attached metal boxes to the studs where we had marked the location of jacks and cable connections, then ran the cables back to the utility room. He numbered or lettered each end of every piece of cable and made a master list; activating any location is a snap. After the plastering was completed, he installed the telephone jacks and cable devices and cover plates.

We decided upon the library as the central location for the audio equipment, with speaker locations marked in the kitchen, family room, all-seasons room, living room, dining room, laundry, and bird room. Bobby didn't need metal boxes at these locations. He just stapled the wire to a stud so that it could be brought through the wallboard as we desired.

So far, everything I've activated that Bobby installed wiring for has worked flawlessly. Bravo, Bobby!

Putting On Airs

ALMOST ALL OF the stages of construction I've de-
scribed so far made a visible difference at the site. Even
the rough plumbing, its pipes hidden in floors and partitions, terminates in kitchen and bathroom fixtures;
the wires of the electrical system run along many of the hidden pathways
between the floor joists, studs, and rafters but terminate in brilliant lighting
fixtures. Only the heating system has tried to stay out of sight — though not
out of mind.

On February 19, 1993, Barry Risk visited me at the house, and we sat at
the familiar picnic table, which had already hosted an impressive number of
conferences, meals, and carpentry tasks. It must have been a cold day. I remember regretting not being able to offer him a warm room. The subject was
air-conditioning! I was taking the first step to introduce three invisible systems into the house, each of which affects the quality of air inside. One makes
it cooler and dryer, one makes it fresher, and one makes it comparatively
dust-free.

The trail to Barry Risk, a sales rep for an air-conditioning system, began
with one of our *This Old House* projects, a nineteenth-century house in Way-
land, Massachusetts. An air-conditioning system was "retrofitted" (a retrofit
is a component that was not part of the original construction) as part of the
renovation. Later I visited Richard Trethewey's home, where Richard

showed me the same system retrofitted to the renovated kitchen of his older house. Steve Thomas, my partner on *This Old House,* retrofitted his nineteenth-century house with this system on Richard's recommendation. It should come as no surprise that I was thinking of installing in our *new* house a system designed especially for retrofitting existing houses.

"An air-conditioner is really a giant dehumidifier," Barry says. The system he described to me draws warm-moist humid air from the house through a return duct to an "air handler," where it passes over a six-row-deep refrigeration coil. As it cools, the air sheds its humidity. A high-speed, permanently lubricated motor in the air handler forces the cooled air at high velocity through a six-and-a-half-inch-square "plenum," or duct, as far as possible toward the area of the building to be cooled by the system. When the duct reaches partitions and floors it can't penetrate without major disruption and expense, the delivery system downsizes to flexible two-inch aluminum tubing wrapped with insulation and a foil covering. The total diameter is about four inches, and since the insulation around the tubing can compress, it is relatively easy to snake through the structure to the outlets or "terminators." The three feet of tubing next to the outlet is made of sound-reducing spun nylon to make the cool-air delivery quiet. Since the outlets are only two inches in diameter, they aren't very obvious in a room. Air comes out of the outlet in a jet. (I think the mechanics are something like the systems that deliver fresh air in airplanes, except that the outlets don't hiss the way the swivel-head airplane jets often do.) The jet stirs the air, creating an invisible wake, and then mushrooms to affect all of the air in the room. A person in the room is not aware of a draft from the air-conditioning, but there is enough gentle air motion to make the air temperature pretty constant from floor to ceiling — distinguishing the system from others that create air layers of quite widely varying temperatures.

Barry laid out our house in four separate systems, each with an air handler in an attic space or in the basement, and each with a compressor outside the house with lines less than an inch thick carrying cold coolant from the compressor to the air handler coils, and warmed coolant from the air handler back to the compressor. The system takes about a tenth of the space of conventional air-conditioning installations and removes a third more moisture from the air. The plenums are tucked under the rafters near the outside wall, where they don't even interfere with storage space. Except for the exterior compressors, which are in locations where we can screen them from public view, the air-conditioning system is a marvel of disguise.

The increased dryness of the air transmitted from the air handlers means we can set the thermostats a little higher in the summer and be entirely comfortable. I was preparing for those stretches of hot and humid weather that can be frequent occurrences during the summer here in New England. I expect the air-conditioning to be working regularly from June through September. Barry promises that operating costs will be a little lower than average to offset installation costs (materials more than labor) being a little higher than average. The most cost-efficient way to operate the system, he says, is not to turn it off when we go out but to turn the thermostat up a little. Won't it be nice to come home to a reasonably comfortable house on a scorching day and then turn the thermostat down a few degrees to perfect coolness?

On his first visit in February, Barry gave me material to study on the air-conditioning system he sells, and he gave me some indication of what the system costs to install and operate. While he was there, I brought up the subject of fresh air. The building code in Massachusetts now requires fan-driven ventilation for every new bathroom. It used to be that fans were required only if the bathroom lacked a window, but now code requires a fan even when there is a window. Two things about bathroom fans really irritate me. Even the best ones make too much noise to suit my taste, and they waste energy because they take heated or cooled air and simply dump it outside. Barry's suggestion was to put a ventilating fan in the attic of the main house, connected to all the bathrooms of the main house. This might solve the noise problem, but it still wasn't energy-efficient and there wasn't any convenient place to install the exhaust outlet for such a fan on the eave of the house. Traditionally, house venting has been concerned principally with the removal of stale air from bathrooms and kitchens. I was concerned with the quality of air in our entire house. The outer skin of the house, particularly with its solid layer of rigid insulation, was relatively airtight; the house wasn't going to breathe the way houses used to, through a network of small openings, cracks, and crevices. Besides, I was convinced that I wanted more than a conventional venting system. I wanted to consider an energy-efficient air exchange system that would regularly discharge stale air from the whole house and replace it with fresh air drawn from outdoors.

One inquiry led to another. In April of 1993, Tony Joyce visited the house and sat at the same picnic table where I had talked to Barry. Tony is an engineer and salesman for a company about a twenty-minute drive away that sells and installs mechanical systems including air exchangers. Tony studied the plans with me, walked through the frame, and went away and developed a proposal for an air exchange and venting system. I liked what he proposed but saw that the way I wanted the system to function would be a challenge for my electrical contractor, Bob Russell, who was waiting for my final decision about the air exchange and venting system so that he could complete the rough wiring in the bathrooms. Tony and Bob conferred on how to control the system to provide both the air exchange and the bathroom ventilation with essentially one central piece of equipment accomplishing both purposes.

The utility room didn't have much available space after the installation of the heating equipment, so Tony located the air exchanger in the storage area at the back of the garage. It was about as far as it could be, within the footprint of the house, from the three bathrooms on the second floor of the main house, so Tony advised mounting a second fan in front of the exchanger to help pull air from the main house. This fan had to be carefully balanced with the fan in the air exchanger so that it didn't either starve the main fan by providing it with too little air to exchange or stress it by pushing too much air into it. The air exchanger vents stale air out through the west wall of the garage, and draws in fresh air through the north wall of the garage.

Branch ventilator lines from the second-floor bathrooms feed into a seven-inch trunk line in the duct system. The system is activated each time a bathroom light is turned on, and switches off when the light is turned off. As the trunk line comes down the stack of pipes and ducts behind the curving wall of the stairwell, it picks up the branch ventilating line from the kitchen. In the basement, the trunk line picks up the branch line from the first-floor powder room and then comes across the basement toward the garage; it picks up a final branch line from the bird room before reaching the air exchanger. In addition to being activated by specific activity such as the turning on of a bathroom light, the air exchanger is activated by a timed control to operate for twenty minutes every two hours to refresh the air in the house. This means that even when no one is home, the exchanger can do its job refreshing the air in the house. There is a control mechanism in the utility room where this timing can be set to our wishes, and the speed of the fan adjusted.

One feature of an air exchanger that I admire is that it addresses energy conservation. During the winter it not only exhausts stale air but it captures a portion of the heat in that air, transferring it to the cold incoming air, so that the heating system doesn't have to use very much energy to heat the fresh air. Our exchanger has a disk feature, which transfers about 80 percent of the heat in air being dispelled to the incoming air without mixing the two volumes of air. As fresh air comes into the house, we have routed it into the return duct for the air-conditioning system for the great room in the ell. In the winter it simply migrates to other areas of the house. In the summer the same thing occurs, except when the air-conditioner for the ell and all-seasons room is running, when it is drawn through the air handler and dehumidified before it enters the living space.

Our fresh-air system is even quieter than our cool-air system. What, then, of the third air quality control — keeping the environment relatively dust-free? It, too, is a comparatively quiet system. In our old garrison colonial, we enjoyed the benefits of a central vacuum system. It wasn't part of the original construction but had been retrofitted by the previous owner. Amenities like that become part of one's way of life. Laura and I reaffirmed in very early discussions of the new house that it should have a central vacuum system.

When our central vacuum system is on, there is no noise in the house from a vacuum cleaner motor. The only noise is the hiss of air being drawn through the attachment on the flexible plastic hose. There are no bags to deal with, no filters to clean, no heavy tank to pull. The system is so simple that Joe Manzi's one-man crew installed it in less than two days — one day to rough in the outlets, piping, and control wires before the drywall was installed and a few

March 1993. The exterior awaits clapboards and trim; most of the construction activity is inside the structure.

hours to install the vacuum unit and the outlet covers after the house was plastered. There are eight outlets placed strategically throughout the house. Each is a "flapper," like a typical outdoor electrical outlet. You have only to lift the flap and plug in the end of the thirty-foot hose to activate the system; as soon as you unlink the hose, the system shuts down. (We've found that the vacuum system works very well even when two different outlets are activated at the same time.) There is an outlet in the second-floor hall and one in the master bedroom of the main house to service the bedrooms and bathrooms; a second in the first-floor hallway to service the living room, dining room, and library; a third in the all-seasons room that reaches into the kitchen; a fourth in the peninsula between the kitchen and family room to vacuum the west half of the ell; a fifth in the mud hall of the office wing; a sixth in the hall outside my office that reaches it and the bird room; a seventh right at the power unit, which, like the air exchanger, is in the storage area at the rear of the garage; and, finally, an outlet near the garage doors so that we can vacuum out the cars either inside the garage or on the driveway close to the garage.

The piping for the central vacuum consists of two-inch PVC pipe, but much thinner than the PVC used in plumbing construction. Sections of the pipe are sealed with PVC cement. Making the pipework airtight is obviously essential. The straighter the pipework, the better — to avoid jams — but the thin pipe does come with elbows and other standard fittings as may be necessary. The container into which dust and other small debris is deposited at the power unit has about a three- or four-gallon capacity. It is clear plastic so that the need for emptying can be easily checked. The air sucked through the pipework is vented through the garage wall after the waste material in it is screened into the plastic holder.

In my experience, these systems are more powerful than portable vacuum cleaners. The accessories are more sophisticated than those we first used in the central vacuum system in the old house. Attachments for cleaning carpets then had a separate power unit. In our new installation, the beater bars of the carpet cleaner run off a little turbine powered by the air suction itself. Laura and I made sure we provided in the new house what we thought valuably convenient in the old house: two sets of hoses and accessories, one for the second floor, one for the rest of the house, which is more or less on the same level. In fact, we have a third hose attachment — the backup hose we bought for the old house — to keep in the garage for cleaning the cars.

I think the initial cost of installing our central vacuum system was about three times as much as buying two top-of-the-line portable vacuum cleaners. But remember how frequently vacuum cleaners have to be serviced and rebuilt. Our old central vacuum system ran for a decade without any maintenance. I expect our new installation to run a long time without problems.

With these three systems to provide good air in our house, the story lies more in the technology than in the installation. The key figures are the engineer and salesmen, who adapt the systems to the plans of your home. Installation has to be done carefully, of course, and in our case I was pleased with the installations.

The air handlers and fiberglass duct board plenums for the air-conditioning arrived on March 30, 1993, the same day we took delivery of the exterior window frames for the main house, and the front door. From then until the plasterers came in August, technicians were in and out of the house on a patchwork schedule installing the three air systems. Four different firms contributed to the installation of the air-conditioning alone. I was in no hurry. They could take as much time as they wanted provided they completed their work before the drywall hangers covered the interior walls with blueboard.

Home Cooking

In 1987 I saw a kitchen in a very old farmhouse Tommy Silva was renovating. We videotaped a "remote" (a short feature shot at a different location from our main renovation project) there to show how a technologically modern kitchen can be designed to look at home in a period renovation; the house we were featuring on *This Old House* at the time was also an old farmhouse, dating from the eighteenth century. The new kitchen cabinets appealed to me at first glance. The closer I looked at them — the wood, the finish, the construction, the pulls — the more impressed I was. The cabinets had been made by a company in Bath, Maine, that specializes in period-style cabinetry. Tommy couldn't say enough about the quality of the cabinets and the ease of installation.

In late March of 1993, I flew to Philadelphia to make a personal appearance at a home show. The Bath, Maine, company had a booth there. One of their reps came over to chat with me during a break between my presentations. We recalled the televised kitchen, and the rep remarked how much business they had derived from the exposure of their cabinetry on *This Old House*. Jokingly I said, "I'm building a house for my family now. You're probably too busy to fit a kitchen for us into your schedule."

By mid-May, I wasn't joking any longer. Our schedule was in disarray. In April, what had seemed to be a heavy cold or chest infection that Laura suffered for several days without improvement turned into pneumonia. When she consulted a physician, he immediately sent her to the hospital. For a few weeks, no one paid much attention to the new house. The illness left Laura very weak for several weeks. More of my time than usual was devoted to helping manage our household. Yet we still hadn't abandoned our intentions of moving into our new house in the fall of that year.

Reluctantly I acknowledged that I didn't have time to build the integrated kitchen and family room cabinetry. If I couldn't do it myself, I had known since 1987 whom I would choose to do it for me. I telephoned David Leonard, the co-owner of the company, fully prepared to hear him say that he'd like to build the cabinetry for our house but that I had waited too long for him to be able to schedule us in mid-1993. He said they were very busy, but that he wanted to come down and talk with Laura and me at the new house; we'd talk schedules later.

Laura and I had done our homework well before Dave's visit. The choice of appliances for a kitchen drives some of the design. Laura had assembled a lot of appliance catalogs, which she and I studied so as to be able to specify the appliances to Dave early in our talk.

Dishwashers are usually not a design problem. Most of them are of a standard size — twenty-four inches in width with height that fits a thirty-four-and-a-half-inch opening. Most countertops are thirty-six inches high, so the standard height of a dishwasher allows for an inch and a half of counter over it. We chose a top-of-the-line dishwasher, not for the bells and whistles, which we largely ignore, but for the promise that it operates very quietly. It has kept its promise. Its operation is not bothersome even when we're standing next to it.

The refrigerator was a big issue because it is such a big appliance. We wanted it to be as unobtrusive as possible. The manufacturer we chose makes side-by-side refrigerators and refrigerators with freezer compartments at the bottom but none with a freezer compartment at the top. I thought a bottom freezer compartment would be awkward because the drawer would pull out toward the island close by; side-by-side doors don't take up as much space when open as the bottom drawer does. I'm not crazy about side-by-side's — the narrowness of the compartment makes it harder to organize the refrigerator space, in my view — but it was the best choice in the circumstances. What is more important to me is that the refrigerator is only twenty-four inches deep, so it doesn't jut out as far as the average refrigerator at twenty-eight or thirty inches; and it accepts wood panels on the door to match the other cabinetry in the kitchen. It was in our plans from the beginning to install a backup freezer in the pantry/laundry room, but I think we may install another whole refrigerator/freezer unit — probably with a refrigerator over a bottom freezer drawer.

Choosing a microwave was a simple task. We went with the same make that has served us well for several years in our old house. The old one was portable. The new microwave is installed in the upper cabinetry and is the most prominent appliance in the kitchen. For ovens Laura wanted basic, black, self-cleaning, and two of them. Many times over the years she has found herself needing a second oven. But she declined having elaborate controls on the ovens.

For a cooktop we chose a four-burner appliance with the grill and griddle accessories that are understandably popular these days. It has a very clean contemporary look. It deals with cooking vapors by pulling them down through a grill with a downdraft fan rather than venting them up under a hood, enabling us to avoid hoods, which I find unattractive.

The final appliance chosen was a garbage disposal. Usually septic systems don't tolerate garbage disposals, but our septic system is engineered to accept kitchen waste after it is processed in the disposal. I think we shall be cautious about what we feed it until we build up a level of confidence in the engineer's promise.

Dave Leonard didn't dive right into design questions when he came to the house in late May. He first asked Laura and me a lot of questions about how the great room of the ell would be used. Where were the main traffic lines going to be as people crisscrossed it going from one part of the house to another and moved about the great room to cook in the kitchen area, eat in the dining area, watch television, or sit by the fire in the family room? The design and furnishing of the great room must leave those basic paths unobstructed, he counseled, or the family would experience something frustrating about the room even if the reason for the frustration wasn't consciously understood.

How much space should we apportion to the family room area, how much to the dining area (how many did we want to be able to seat there?), and how much to the kitchen? In such a discussion, as Dave says, "you inevitably chip away at dreams a little." Yet we didn't have to cut back too much from our list of all the things we wanted to be able to do in that one open space.

The thing I most appreciated about Dave was his determination to keep Laura centrally involved in the design of the great room, especially of the kitchen area. He wanted it to be *her* kitchen. Even when she was reticent about expressing her opinion, Dave was quietly and politely insistent that she say what she preferred. In the end, his sensitivity resulted in some thoughtful details. In our old house, Laura preferred to do baking on the kitchen table, which was lower than the kitchen counters. The table height was more comfortable for her. There wasn't space for a table in the middle of the new kitchen, but Dave designed a baking center in the peninsula that separates the kitchen from the family room. At the baking center, the counter drops down from thirty-six inches to thirty inches, and the surface is polished granite, a great surface for a pastry chef.

Dave also ascertained that I'm not home nearly as much as Laura and Lindsey would like and that Laura wants to have as much contact as possible with me when she is working in the kitchen. To make it comfortable for the family to sit in the kitchen, Dave designed a sitting place on one side of the island. He also persuaded us to lower the height of the peninsula between the kitchen and the family room.

Our first take on this peninsula when we discussed the kitchen with Jock Gifford was that the peninsula should be about eight feet high, so that oven space, cabinet space, and space for the return duct of the air-conditioning system could be incorporated into it. But a high peninsula effectively cut off sight lines between kitchen and family room and threatened to diminish appreciation of the timber- frame architecture of the room. "We'll find some way to tuck the utilities in," Dave said, "but let's cut the peninsula down so that its highest surface is a forty-two-inch-high ledge on the family room side of the peninsula." The counter on the kitchen side of the peninsula sits lower at thirty-six inches. Because of that slightly higher ledge, a person in the family room will not be distracted by seeing utensils on the working surfaces in the kitchen.

Jock Gifford's Design of Our Kitchen

Our conversations with Dave modified but did not fundamentally change the design of the kitchen Jock Gifford prepared for Laura and me when he designed the house, and that we were very comfortable with. The kitchen is laid out in a squared-off U with an island located in the open end of the U. The refrigerator is on the right wall as one looks into the U. The sink and cooktop are both located on the wall at the base or far end of the U. Kitchen designers always analyze the efficiency of the triangle linking refrigerator to sink to stove when they appraise how well a kitchen works; in our case, the triangle is reasonably compact.

When we accepted Dave's suggestion to lower the peninsula, we left ourselves the problem of relocating the wall ovens. He provided the solution. Tuck them into the north or inner side of the island side by side, he proposed. From a design standpoint, it is a good solution. When you walk into the kitchen, you don't see a lot of black appliances staring you in the face. This placement brings the ovens down below counter height, but Laura hasn't minded their height.

In June, Laura, Lindsey, and I drove to Bath for another consultation with

Dave and to tour his shop. It appeared that he meant to work our cabinetry into his schedule. I was eager to see this shop, which replicates, as far as possible and reasonable, the woods and hardware, construction techniques, and finishes of the period or style of house for which the cabinetry is being designed and built. Much of his work involves cabinetry for old houses that are being renovated to make them appropriate for contemporary lifestyles.

Houses in the colonial period, of course, did not have the kinds of kitchen cabinetry that Americans take for granted as essential in their kitchens today. Colonial kitchens had furniture — tables, chairs, racks, free-standing cupboards, and so on — but not walls of built-in cabinets. Dave's company's way of dealing with this historical irony is to design period kitchens in which the cabinetry has a "furniture look" to it.

Our conversations in Bath dealt with details of the kitchen and then moved to a discussion of matching cabinetry for the family room next to the kitchen. For each side of the great stone fireplace in the family room Dave designed bookcases. In the northwest corner of the family area, the bookcase flanks a large corner cupboard to hold a television set, a videocassette player, and a stereo system. We discussed whether to have swinging or retractable doors on the cupboard. I argued for swinging doors. I thought that if the doors were retractable, they'd be open all the time; but if they only swung open, we would make the effort to keep them closed when the equipment inside was not in use. I know television provides my livelihood, but I still prefer to keep the set behind doors when not in use.

Next to the corner cupboard on the north wall is a large casement window. Dave designed cabinets to fit around it with a window seat underneath to make the window almost become part of the cabinetry. Between the casement window and the door out to the north terrace Dave found room for a writing desk, a secretary built into the wall, a nice place, he said, "to hide the bills."

What the proposed sequence of bookcases, cupboard, window seat, desk, and kitchen cabinets amounts to is a sweep of woodwork, interrupted only by the fireplace and an exterior door, on three walls of the great room. I suggested cherry wood for all of it. Cherry wasn't used abundantly in the colonial period, but I like the grain, the color, and the subtle elegance of it. The craftsmen whom I met and swapped woodworking lore with in Maine made all of

A woodworker in the shop where our kitchen and family room cabinets were made shows the solid wood and dovetail-joint drawer construction, which are signs of quality.

our cabinetry with three-quarter-inch solid cherry stock. All of the doors have raised panels. The pulls for drawers and doors are of solid cherry turned by hand on a lathe. The hinges are very simple nonmortised hinges imported from England; only the bronze barrel shows.

The craftsmen of Dave's company take great pride in their finishes. The interiors of the cabinets were painted — three coats — in Coldwell green, using a semi-gloss paint. Then the exterior surfaces were finished in a process that contains a few trade secrets. The intent is to make the wood look of a period, but not "antiqued" in an obvious way. All of the finish is applied by hand. A base coat of stain is brushed on. Several coats of stain and varnish are added, with drying periods in between, to build up a patina. There are no shortcuts. Lacquer finishes can be completed in a matter of hours, but the non-lacquer process in Bath takes a week. "A person seeing the finish for the first time," Dave says, "knows there's something different about it but he can't put his finger on it." Wood treated this way, Dave promises, ages handsomely. His most beautiful kitchens, he says, are at least ten to fifteen years old. The wood keeps looking better and better at an age when other wood cabinetry is starting to look tired.

Since Dave's customers are scattered over the entire country, installation

If we hadn't had double doors, we might not have gotten this shopmade corner cupboard into the house.

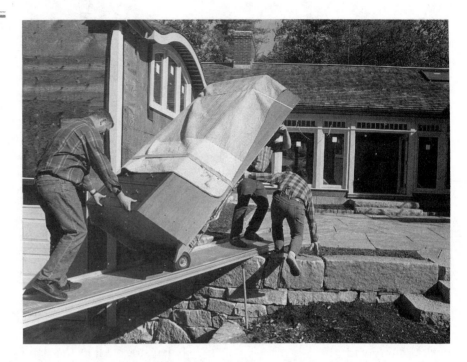

of the company's cabinetry is generally not done by the firm itself. What the craftsmen do is measure and remeasure everything very carefully. They are the very personification of my motto "measure twice, cut once." Then they send off the finished cabinetry with a detailed installation manual, commending the installation to other hands.

All the woodwork for the kitchen and family room arrived in one shipment in late August 1993. My dad and I installed it, a partial consolation to me for not having had a chance to build any of it. What a pleasure it was to handle such beautifully made pieces! The company anticipates every need. They allow extra wood on edges that show, giving the installer flexibility in making the cabinets perfectly plumb if the wall they're being attached to is slightly "out," or off-plumb. They send scribe material to facilitate scribing the pieces to make them fit closely against uneven abutting surfaces, for example, where we fitted bookcases against the uneven stonework of the family room fireplace.

We started on the kitchen first because I had to get the lower cabinets installed before countertops could be set. Working with an old pro like my

Even though I didn't make our cabinets, I did install them, and who better to join me than my father?

father ensured that the installation would go smoothly. The upper cabinets were extremely difficult to maneuver into place without the coordination of two carpenters. The job actually required two men and an electrician. Bob Russell was in and out several times to install wiring and switches, and put electrical boxes on the island. I drilled holes on the underside of the upper cabinets so Bob could install the "puck lights," which illuminate the work counters beneath the cabinets. I installed the appliances as we got the cabinets for or around them installed, and Bob completed the wiring of them.

Basically, the cabinetry installation fell into two phases. When the pieces arrived, my father and I worked as often as we could to get them installed but without several kinds of trim in place — kickboards, shoe molding, or cornices. Kickboards, for example, had to be put aside until the tile floor was installed in the kitchen. There wasn't any rush about any of the trim. I left it all for a later day in order to concentrate on the fundamentals remaining between us and occupancy.

A couple of the upper cabinets to the right of the kitchen window could not be installed until the countertops were set over the lower cabinets. At one place the upper cabinet comes all the way down to the countertop. The bottom section of the cabinet is a door or flap that pulls up to reveal a good-sized space I call the "appliance garage." The "garage" floor is the countertop. It is meant to house such things as a toaster, a food processor, and a blender; they are immediately at hand but not sitting out all the time.

Of all the counter materials on the market, the most widely used is plastic laminate: a series of thin layers bonded together under pressure and heat; the outer layer is plastic-bonded to a core of particleboard. Laura and I chose solid-surface counters instead. They are slabs of acrylic-based material, seamless, and non-laminate. The kitchen sinks and counters extending out from the sink are all one unbroken piece of solid surfacing. The material is more expensive than the laminates, less expensive than such materials as granite or marble. It stands up to ordinary wear and tear well, resists cuts and burns, and can be repaired with filler in the event of a truly disfiguring accident.

I had trouble getting the right piece of polished granite for the countertop of Laura's baking center. Just this one minor detail frustrated me for weeks. I found exactly what I wanted at a stone dealer's, and they said I could have it. But later they said they couldn't find it. Their attitude seemed to be that if they didn't have what I wanted, I should select something they did have on hand. It was too much bother for them to acquire a single small piece of what I wanted. Finally, I gave up on the dealer and went to my tile men, the Ferrante brothers. Yes, they could get it for me. But a scheduled delivery didn't occur. There was an excuse, then a delay, and, finally, when I thought I would bite completely through my lip in annoyance, a delivery.

When the kitchen was completed — all the cabinetry, appliances, wiring, and countertops in place — it looked handsome enough to photograph for a magazine. I found it hard to believe Dave Leonard's claim that the cabinetry will look even better ten or fifteen years from now. It was already a dream kitchen.

The kitchen is complete except for installation of tile backsplashes above the sink and behind the cooktop. It is everything a contemporary kitchen should be. On the right, you can see a thin slice of the plywood-and–two-by-four stair railing in the front hall — one of those unfinished items awaiting my attention and raising Laura's reasonable apprehension about how much finish carpentry will remain unfinished and for how long.

A Working Bathroom

IN THE VERY FIRST stages of building our house, when we were clearing the site, roughing in a driveway, and excavating and blasting for the foundation, I recall that we fretted more about the septic system than about any other aspect of construction. It appeared at one point that we might have to build up the septic field with sandy soil near the intersection of the driveway and the access road coming in from the main subdivision street; but if we raised the septic field, we might also have to raise the entire driveway. At another point, it appeared that the elevations weren't going to permit the flow of waste from house to septic tank to leaching field by gravitational flow — in which case we would have to induce the flow with an expensive pump, with reliability something less than the force of gravity's.

Herb Brockert solved this problem for me. On April 28, 1992, almost the last possible day permitted in the spring to test the water table (while the table is at its highest and before the dry season depresses it and makes tests unreliable), Herb suggested we redo the water table test done in 1985, using his heavier excavating machinery, which could more easily dig through the rocky subsoil. We suspected that the original test had been performed with a much smaller machine and that when the tester hit the first large boulder he quit because he had gone deep enough to build a proper septic field and he had not hit water yet. Herb and I wanted to reach the actual water table to determine how low we could place the field. His excavation showed the water table to be low enough even in a wet spring to give us the depth we needed to install a leaching field without building it up above existing ground level. Our engineer revised the plan. For a second time we petitioned the Board of Health for approval of a revised plan. The board gave its approval on June 9, 1992. We didn't pay attention to the septic system again until June of 1993.

After the design engineer set stakes to locate the beginning and end points and the centerline of each of the three required trenches, I called Herb back to the job to begin constructing the septic system. He stripped off the topsoil of the septic field with his excavater to get down to the basic subsoil of the area. According to the plan, he had to remove all of the subsoil under and within ten feet of the trenches. (This excavated material did not go to waste; it was perfect fill for grading around the house.) Herb replaced the excavated subsoil with a soil mixture specified by Board of Health regulations. The health inspector rejected the first sample of replacement material Herb submitted but accepted the second sample, which Herb found at another nearby source. Thus we passed the first of several inspections made during the installation of the system.

With his backhoe, Herb dug three trenches, each about fifty feet long, in the new subsoil he had installed. He placed a foot of three-quarter-inch to inch-and-a-half double-washed stone in the bottom of each trench, on which he laid four-inch perforated PVC pipe. He surrounded the pipe with more of the crushed stone. On the uphill end of the trenches toward the house, we installed a small concrete box called a "distribution box." It has a removable top. There were knockouts in the sides, which could be removed to insert the ends of pipes. Unperforated four-inch PVC pipe connects the box to the pipes in the trenches.

Over the existing crushed stone in the trenches, Herb laid a two-inch layer of eighth- to quarter-inch peastone. Then he backfilled and graded the trenches with some of the material he had excavated to form them. Finally, he covered the entire septic field with the original topsoil. The field was complete but not yet very attractive. Roger Cook took care of the cosmetic problem. He came back in the fall of 1993, planted a four-foot-wide strip of grass along the line where the septic field adjoins the access drive,

Herb Brockert and his assistant, Steve, prepare to close and cover the distribution box, which channels liquid waste flowing by gravity from the septic tank into one or another of the three visible trenches of the septic field.

and covered the entire field with wood chips to make it blend in with its natural surroundings.

Herb installed the septic tank on June 28. The fifteen-hundred-gallon capacity we wanted mandated a rectangular concrete tank, which measured about ten by six by six feet. It was delivered on a truck enhanced with its own built-in crane, so the placement of the tank in the excavated hole was not a difficult operation. Because I had raised the final elevation of the house about two feet higher than we first thought it would sit, Herb was able to set the septic tank higher than its planned elevation and still have gravity flow from the house out to the tank. With the tank in place, Herb and I recalculated the grade from the tank to the septic field and found, to my delight, that there was enough difference in grade for that part of the system to work by gravitational flow also: Did this mean the elimination of the pump chamber? The decision was in the hands of the health inspector. Herb called him out to the site for a firsthand look. The inspector was very reasonable about the new development. He knew that the elimination of the pump chamber gave me a more dependable system. He gave it his stamp of approval, providing we installed a cleanout where the pipe changes direction to detour around the rocky knoll on the way from tank to field. A very small concession to make for a major simplification of the system. What a relief!

Just when we thought there was clear sailing to complete the whole system, Herb discovered subsurface granite blocking part of the path where he wanted to lay the underground pipe from the tank down to the distribution box. Back came the same blasting crew that had blasted the rock in the way of our basement. This time it took them less than a day to complete the blasting. The path of the trench was by no means the shortest line between tank and field. To avoid the knoll, the old foundation, and trees we wanted to save, the engineer plotted the trench as straight as possible from the tank in the front yard down to the driveway, then along the curving driveway to the septic field.

The blasting fascinated Herb, but I was nervous. Most of the blasting occurred within thirty or forty feet of the facade of the house. What if the mat didn't contain the debris from a blast? Would I be replacing some broken windows at more than $100 a sash? Would I have to replace scarred siding? When they shot the charge I had my fingers crossed. Some debris did fly but not in the direction of the house.

After all the parts of the system were installed, the design engineer returned to obtain measurements and grades so that he could submit for certification not the plan that had been approved but the system the way we actually constructed it. The final approval came back without a hitch.

With the septic system operational, I could hardly wait to get a plumber in to hook up one of the bathrooms. We elected the powder room on the first floor of the main house as the most centrally located bathroom for anyone working in the house. I had our plumber install an inexpensive, out-of-date toilet I purchased to use until the floor was tiled and the final carpentry completed — at which time we would have the final fixtures installed. For quite a while only a shower curtain hung in the doorway to give privacy, but the plumbing worked great. I know that the septic system is the least romantic system in the house, but after a couple of years of relying on a portable toilet standing at the edge of the driveway, the first inside straight flush was royal music to my ears.

Lucking Out

IT'S NOT SURPRISING that we think of unexpected problems in construction as the beginning of regretted compromises. Most of the time that is the case, but not always. Not long after we began the installation of the windows, doors, siding, and exterior wood trim on our house, I experienced a very happy outcome after discovery of what at first appeared to be a perplexing problem.

The windows for the main house had arrived and been installed in late November of 1992. John Conrad and Bill Delaney from our framing crew returned to help with part of the installation, and my son Bobby helped me with the rest of these windows. The double-hung windows are state-of-the-art — wood on the inside frame for appearance, aluminum cladding on the outside for easy maintenance. The three-quarter-inch-thick insulated glass contains a nearly invisible reflective film suspended between its two sheets; the film reflects heat back inside in cold weather, and back outside in hot weather, and it filters out much of the ultraviolet rays from the environment, which cause deterioration of carpets, wood, fabrics, art, and other materials inside the house.

As received from the manufacturer, the windows were as unfinished visually as they were complete and sophisticated technologically. Nailing fins were attached around the outside of the frames. Before setting the windows into the rough openings, we attached nine-inch-wide felt paper splines onto the plywood next to the sides of each opening. Then we set the windows in the openings from the outside. After adjusting the placement to make the windows plumb and square, we drove roofing nails through the fins to secure the windows. At this point, the windows filled the openings, but they looked naked. The manufacturer had left the design and application of exterior casings and trim in the hands of the builder.

For the main house and the south gable of the office wing, I wanted to do something simple but special in the way of trim details. (For other windows on the west side of the house, I could make and install sills and flat casings myself.) Jock Gifford sent me sketches of a simple and historically authentic plank frame that he found on an old house on Nantucket and has since replicated for other houses he has designed. Laura and I liked the look of it very much. I authorized Bruce Killen, a Nantucket woodworker, to make up our plank frames.

Bruce made up the frames in recycled redwood. The fit has to be excellent, and he is meticulous in his craftsmanship. But there was a small error either in the measurements I transmitted or in the measurements Bruce used in his first model. When the frames arrived for me to install, the opening of each one was wider than it should be to look right. There was too much of a gap between the window and the frame. It wasn't a big difference in inches, but it made a big difference to me as I looked at the first frame I installed.

There was no alternative, in my mind, to shipping all the frames back to Nantucket for alteration. These were not casually assembled frames, however. Each one was fastened together with adhesive and screws, and then the counterbored holes for the screws were plugged. Word came back to me from Nantucket: it wasn't feasible to take the frames apart, recut, and reassemble them. They had been made too well in the first place. They would be ruined in the disassembly.

Bruce studied the situation for a few days. He proposed adding a molding on the inside of the frames that would slightly overlap the aluminum cladding of the window unit. When I saw a sketch of what he proposed, I thought it looked even better than the original treatment! From the jaws of an unin-

tended but potentially expensive error of his or mine, Bruce seized an inexpensive solution that was also an improvement. Talk about lucking out.

The window frames with their new moldings came back to the house in the same delivery with the front entrance unit in which a mahogany door was framed with a row of transom lights (rectangular panes of glass) above it and vertical rows of side lights. I had wanted very much to make the front door myself, but again there wasn't time to do it in the time period when I needed to get the door finished and installed. After Bruce had primed window frames and I had applied one coat of stain, I secured the frames to the frame of the house with six four-inch rust-resistant screws per window. Plugging the holes counterbored for the screws completed each installation. I used both screws and nails in the installation of the front door and gave the wood a coat of primer to protect it.

Before I finished trimming the main house windows, the custom windows arrived from Connecticut for the window walls of the ell and all-seasons

After I check the levelness of shims in the opening (opposite page), we lift an eleven-foot-long section of ell window into place. It weighed several hundred pounds. We carefully slid or skidded it from the delivery truck across the terrace over sheets of plywood. This was a window I didn't want to break during installation!

The eyebrow-like window approaches its rough frame. Since I had done all of the framing for it, my framing skill was at stake. It slid in, with no room to spare.

room, and for the eyebrow-like window in my office. The manufacturer sent everything necessary for the installation, including both interior and exterior casings. The wood of the exteriors had all been sprayed with a very tough epoxy paint matching the color of stain we intended to use for all of the exterior trim on the house.

The weight of the insulating glass and the mahogany frames of the large units for the window walls made them much more formidable to unload, position, and install than the double-hung windows. For the largest units it required the combined strength of the delivery man, John Conrad, Bill Delaney, Bobby, and me to get them into place. They were beautifully made. There was no way we could have achieved that look from standard parts.

With windows and doors very near to complete installation, we lacked only garage doors to have all the house openings closed in. By the time I investigated various garage door styles and prices, ordered the doors, and got on the schedule for installation, July of 1993 was upon us.

Garage doors come in wood, fiberglass, and steel, but in my view the best insulated ones are steel. Since I had put some radiant heating tubing in the concrete floor of the garage so that I could heat the garage modestly in very cold weather, I didn't want to lose any more heat than necessary through the doors. Within a few miles of our house are some eye-catching old New England barns with a row of transom lights above their great doors. I had hoped to replicate that look by installing transom lights above the garage doors. The way the elevations worked out, however, there wasn't enough height to make that detail feasible. The closest I could come was finding a steel garage door with a horizontal set of windows in it. I elected to have the glass in the top row of panels of each of the two doors. It wasn't transom lights, but it wasn't bad.

The garage doors I liked came in two possible colors — white and off-white. I asked to see paint chips. The off-white chip was an almost perfect match for the color we had chosen for all of the exterior wood trim on the house. The doors wouldn't have to be repainted. Talk about lucking out! I had already done it twice on the exterior details of the house.

Because of the size garage doors I ordered, it took an extra week to get delivery of them. In our old house, the garage door is ten feet wide. Most standard doors for residential garages are nine feet wide and seven feet high. I had come to appreciate the extra width of the ten-foot door for parking our Bronco with its wide side-view mirrors on both sides. If I wanted ten-foot doors, I had to wait another week. During that time I installed the jambs and casings for the twin doors. A garage door specialist came and installed the doors and the automatic opening system.

From the time I first thought about building a new house, I knew without doubt that the siding would be clapboard. I selected half-by-six-inch clear red cedar clapboards delivered already primed for stain on all sides with a primer that prevents tannic acids in the cedar from bleeding through the stain. As I mentioned in the opening section of this book when I was speculating on the reasons nails were popping on the siding of a house I was evaluating, some builders choose not to stain or paint the side that can't be seen. But I believe strongly that the unseen side should be treated to encourage the wood to adapt uniformly to moisture whether the moisture is attacking the outside of the board or the hidden side.

Jock Gifford showed me a sample of clapboard with a half-inch, half-round decorative bead along the face of the butt edge. He found this ornamental feature on some very old houses on Nantucket. Laura and I really liked the beading but not for every side of the house. We did what some builders have done for hundreds of years. We used the beaded clapboard just on the front of the main house. There is enough demand for this item that the supplier could deliver it already beaded; it didn't have to be custom cut. The clapboards come with one side milled a little smoother than the other, and the builder can elect which surface to expose, which to hide. I elected to expose the smoother side because I thought it would make the house look more like an old New England colonial.

For the wood trim around the clapboard, I ordered a lot of five-quarter cedar. "Five-quarter" means that the boards were rough cut approximately an inch and a quarter thick; by the time they are dried and planed they are between an inch and an inch and an eighth thick. I bought the cedar trim from the same supplier where I got the clapboards, but the wood for trim was not primed as delivered. Bobby and I primed all four sides of the trim boards with the same primer that had been used on the clapboards.

I wanted good thick boards for the corners, the water table (the horizontal piece of trim that runs around the house just above the granite veneer at the top of the foundation), and any trim around the windows where the clapboard was going to butt up against the trim; if I used only three-quarter-inch boards for trim, for example, the butt edges of the clapboards would project beyond the face of the trim. Because of time pressures, I had earlier delegated to John Conrad and his framing crew the job of installing exterior trim around the roof so that the roofers could come in to lay shingles. But now I happily took over the task of installing the rest of the exterior trim myself.

First, I installed vertical corner boards on all outside corners to provide a transition between the clapboards meeting at the corner; otherwise each course of clapboards has to be mitered where they come together at corners. At inside corners I used a piece of inch-square cedar. I prepped each corner of the house by fastening an eighteen-inch-wide strip of fifteen-pound felt paper to the plywood sheathing, wrapping the paper around the corner seam. It's an extra step to prevent any infiltration of water through the seam where

My father, my son Bobby, and I install rosin paper before clapboards on the hottest day of July. The paper might have stuck to the plywood sheathing just from our sweat, but we added some staples for good measure.

The framing crew worked mostly without it, but Bobby, my father, and I found pipe scaffolding indispensable as we installed clapboards on a gable end of the main house.

the corner boards meet and then through the seam underneath where the plywood sheathing sheets meet.

From the corner boards I moved on to the water table, once a way of diverting water from the foundation, now more of an aesthetic detail than anything else. All of the trim — and then all of the siding — was attached with stainless steel ring-shank nails (nails with grooves around their shanks to increase holding power) so that the nails could be counted on not to bleed, discolor the wood, or pull out of the wood.

The real trick in siding is getting the trim correctly installed. Putting on the siding is the easier of the two stages, but on the house we were building it was not a small task. It was a family project. My father and Bobby spent a lot of days on the scaffolding installing siding with me. Laura had taken responsibility for selecting the color of the stain for the siding. She looked at many samples before finally choosing a medium tan that looks right for the natural setting of the house and complements the stonework (walls, terraces, walks) that is all about the house.

The "siding crew" worked off and on for a long time as schedules permitted. Our method was to set scaffolding for a section of the exterior, apply the paper barrier to the plywood sheathing, apply the primed clapboards, install any remaining trim and moldings, stain everything with two coats of stain, and move the scaffolding on to the next section. I don't put tar paper undersiding. It creates a moisture barrier unwanted there. I chose to apply red rosin paper, which comes in rolls. It is a very traditional layer to use between siding and sheathing. It provides some separation between the clapboards and the plywood. It covers any holes in the plywood and the seams in the plywood sheathing. It offers some resistance to wind that has worked its way between clapboards and is trying to penetrate the interior shell of the house.

There isn't really any consensus on the question of what barrier to install between the clapboards and the plywood. There are more high-tech air infiltration wraps available. If I were renovating an old house whose sub-walls were composed of boards with many seams rather than sheets of plywood with relatively few seams, I probably would go with one of the air infiltration wraps. But many builders I've talked with don't think these more expensive liners are necessary over plywood.

Where joints occurred between pieces of siding we cut overlapping joints at a forty-five-degree angle. Such joints shed water much better than a joint where two pieces butt each other squarely. They are also less visible because they don't open up the way squared-off joints do when the wood shrinks slightly in dry and hot conditions. Our goal was to install siding with the kind of fit such that anyone looking at the completed and stained surface simply won't notice the vertical seams.

At every joint, and anywhere clapboards meet trim, we placed a bead of latex caulking with silicone before setting the clapboard — to make a waterproof and airtight joint. Originally I had thought to have about four and a half inches of each six-inch clapboard exposed. Later I came to the conclusion that a narrower exposure of about four inches would look better. I had to give some thought to how the courses were going to come out relative to critical lines, such as the tops and bottoms of window trim. In a carefully sided house, the siding courses line up evenly with such items as window and door trim. A four-inch exposure worked out just about perfectly with the other lines.

In July and August of 1993 we installed siding on some brutally hot days. We finished installing and staining the siding of the main house and moved on to the ell and the office wing. As we worked on into the fall, it eventually got too cold to stain the wood as we installed it. One of the days we applied siding — we were working around the eyebrow-like window on the east wall of the office wing — was surely the coldest day of the entire winter! The day after we finished the last strip of clapboards there was a major snowstorm. A few pieces of trim here and there were still missing, but the house was good and tight for the unfolding winter. The task of siding was well enough concluded that the finishing touches became one of those details that await a spare moment; my major attention moved on to other aspects of the construction. It would be the fall of 1994 — a superb golden day in October — before I brushed the last bit of stain onto the south gable end of the office wing.

Lindsey Rents a Room

ONE OF OUR PRINCIPAL reasons for building a house in the town we chose was to make its excellent school system accessible to Lindsey. When the foundation was dug for our house in 1992, I had a construction schedule in mind that would have gotten us into the new house in time for the 1993–94 school year. In September of 1993 Lindsey would enter seventh grade, which in the town where we had been living meant the beginning of junior high school. I thought it was an obvious moment to transfer from one school system to another.

As construction moved along in 1993, however, it became apparent that the house was not going to be ready for occupancy by September. The interior walls were plastered by the end of the summer, but not painted. There were no finished floors yet. Some of the lighting fixtures were in place; many others were not. The only working bathroom fixtures were in the powder room. Fireplaces lacked hearthstones. Everywhere you looked there was something still missing. The house had a very unfinished look to it even though it was largely complete as a structure. Maybe occupancy by Thanksgiving if we were lucky.

I didn't want to accelerate completion of the house in any way that would compromise its quality. All through the spring of 1993 I lessened my nervousness about the crunch between construction schedules and school schedules by reminding myself of a rule we learned about even before we broke ground. If a family was demonstrably going to become resident in our new town during a school year, but hadn't moved in yet when school began, the children could commute from another district until the move-in date. Having a hearing before the school committee to establish the facts of the situation — the reliability of our forthcoming residency in the town — was a reasonable step we calmly anticipated. In the worst case we were prepared to drive Lindsey back and forth from old town to new school for several weeks.

In late spring of 1993 I visited the school I expected Lindsey to enter a few months later. It was a courtesy gesture on my part, but it turned out to be the opening scene of a major family drama. The school doesn't honor the rule you've been counting on, I was told at the principal's office; if Lindsey isn't actually resident in the town, she can't attend the school here. What defines residency? I asked. The student has to sleep in the town at least four nights each week was the reply. I decided there was no point in discussing the irony implicit in defining residency by where one sleeps, not where one works or eats or spends time with friends or even watches television. The school staff defended their position by saying it was a statewide rule.

Without doubt we had a worrisome problem on our hands. The man who measures twice before cutting once is also capable of checking a rule twice. I called the state attorney general's office in Boston. No, they said, there was no state-wide — or, more precisely, *commonwealth*-wide — rule about how to treat admissions to public schools for students who were about to move into a school district but hadn't yet made the move when school opened. Each school district made its own policy about such matters.

A few days later I was at the town hall dealing with another house-related matter, and I mentioned our dilemma to the town clerk. The clerk suggested I contact the head of the school committee and explain the situation in the hope of eliciting a solution. The clerk thought a telephone call might yield a hearing.

I've never had a conversation that ran downhill faster than my talk with the school committee head. She swiftly told me Lindsey wasn't going to be a student in the town until she was a resident of the town. That doesn't make sense to me, I argued, containing my consternation as much as I could. "I'm already a taxpayer in this town. My house is under construction." The school committee head then tried to tell me that I was *not* a taxpayer. "No, you're wrong," I said. "I'm paying taxes on both the land and the house we're building. What you're telling me is that parents who are not residents and not even taxpayers are in a better position than I because they can simply rent a room four nights a week in town for their child or children and get them into the schools here. I think that's outrageous." She didn't much appreciate my remarks, but so far as I could tell I was accurate in my analysis.

When I couldn't get anywhere with the school committee, I went to the town library, asked to see copies of minutes of the school committee's meetings during the preceding few years, and read them for their possible bearing on our situation. It appeared to me that there was some evidence of the rule I had counted on: students who could demonstrate they were going to establish residence in the school district during the year were permitted to begin attending town schools at the begining of an academic year.

But I also read in the minutes about a situation that I believed had paved the way for the committee's intransigence toward Lindsey's situation. A woman who lived in a nearby town but owned land in our town decided she wanted her children to attend school in our town. She engaged an attorney to represent her case, but she didn't win. Reading about the dispute convinced me there was no way I was going to get a concession from the school committee. The last thing I wanted was to get into a legal dispute with the town before I had even moved in; all I would gain from that would be local gossip that Norm Abram was trying to throw his weight around.

In my conversation that went nowhere with the school committee head, she suggested to me that it was not uncommon for families to rent rooms in the town for their children to qualify them for the local schools. She also suggested I might park a trailer on my property as temporary housing. (I knew the Board of Health wouldn't tolerate a trailer on the site as living quarters for a few months.) Laura, at my suggestion, called a couple of local realtors to inquire about rooms we might rent for Lindsey. They were rather disdainfully amused by her request. There's nothing to *rent* in this town, they said; there aren't any rooming houses in our town.

By word of mouth, nevertheless, we did hear of a room available to rent in a house owned by an elderly woman. Laura went to check it out, and it seemed satisfactory. The principal of the school wanted to see a copy of a formal lease, but it was a more informal situation in which we could rent as long as we wished. The owner gave us a receipt month by month for our rent payments. We thought it defensible to interpret the residency rule as much in our interests as we could. Which four nights of the week? How much of each night? Any school days when Lindsey resided at home meant forty miles of driving for Laura twice each day. I gladly bought a new car to make the drive more comfortable, but it was not an easy school year for anyone in the family.

Birdmen

It was early summer — June — of 1993. I had plastering on my mind. All the systems that had to be partially buried in the walls — plumbing, electrical, air-conditioning, air exchange, central vacuum, tele-phone, alarm system, cable television — were in place already or scheduled reliably enough that I knew it was safe to get a plasterer scheduled. My prob-lem was that I didn't have an obvious candidate to offer the subcontract to, provided the price was right.

In my own years as a general contractor I had found one plasterer I liked to use because of the quality of his work, but he had retired from the trade several years before I started building my dream house. Tommy Silva rec-ommended a plasterer he knew. The candidate came and looked at the job and gave me a price that turned out to be by far the highest bid I got. Greg Dana, a wallboard supplier from whom I intended to buy my wallboard, gave me a few names after I mentioned my dilemma to him. I called the most local of them, but the plasterer turned pale when he saw the size of the house and the number of high and complicated areas to be plastered. He said the job was too big for him.

Another plasterer on Greg's list walked though the house with me and gave me a fairly good price. But the way he looked at the job unsettled me. From my perspective, plastering the panels of blueboard between the rafters in the timber-frame ceiling of the great room was going to be one of the most challenging and time-consuming aspects of the job. This candidate looked at the great room, seemed a little perplexed about how to bid it, and then said, "Oh, well, I'll just throw it in with the rest of the job." I didn't want someone to get partway through the job and find he had substantially underbid it, so I reserved judgment on him. I had by now almost exhausted Greg Dana's list, but Gary McKay remained on it. Of all the plasterers I talked to, he made the best impression on me. His price was reasonable, and he seemed confident about achieving the level of workmanship I wanted.

Systems for plastering wood-frame homes have changed dramatically over the years. When plastering first became popular, hand-split wood laths — strips of wood about three-eighths of an inch thick and one and a quarter to one and a half inches wide — were nailed to the studs. Then a thick base coat of plaster was applied over the lath and pressed enough through the spaces separating the laths to hold it firmly in place after it dried. The base coat was built up as desired to a finish surface through the application of more thin coats of plaster. Because of the thickness of the plaster, flat walls and ceilings could be produced even though there were variations in the width of studs or ceiling joints. Anyone who has ever demolished one of those walls knows how heavy the lath-plaster construction was, and how dusty to remove.

By the time my father built his house, the system had evolved into fastening panels of rock lath (a rigid gypsum core with a paper covering) that were sixteen by forty-eight by three-eighths of an inch thick to the studs and then applying about three-eighths of an inch of plaster: a base coat of brown plaster and then a finish coat of white plaster or plaster tinted to the desired color of the room. One drawback of these early plastering systems was that the thick wet plaster released an enormous amount of moisture into the house. It took a lot of time to mix and apply the plaster and then several days for it to dry.

More recently, two materials have dominated interior wall finishes. The first doesn't use wet plaster at all. Sheets of wallboard (a rigid material composed of wood pulp, gypsum, or other materials with a paper covering, and with slightly tapered edges so that the joints can be taped and spackled without creating bumps) are nailed or screwed to the studs. The installer covers the seams with paper tape and a compound referred to as "mud," and then sands and respackles and sands until he has the invisible seam desired. Either paint or a wall covering such as wallpaper is applied as a decorative finish directly to the wallboard.

I chose the other material. Instead of ordinary wallboard I stipulated covering the interior wall frame with half-inch "blueboard," wallboard with a coating of blue paper that facilitates the bonding of plaster to the board. A thin coat of plaster is then applied to the blueboard, giving the desirable finish of a plaster wall but with much faster drying time. Once the blueboard is installed, it is important to get the plaster on; the longer the blue coating is exposed, the less effective it becomes as a bonding agent.

Gary McKay went into construction work right out of high school. He worked concrete at first, but when he saw the difficulty of getting winter concrete work, he made the transition into plastering, first as winter work and later as year-round work after he became expert with it. What sold me on Gary was that he didn't flinch when I told him that I wanted more than just the standard application of a single skim coat of plaster over the blueboard. I wanted him to put a thin base coat of plaster on all the ceilings and then a skim coat over it; and I wanted the ceilings to be finished smooth. The additional base coat helps prevent cracks and makes it more difficult for screws to pop through the finished surface as the wood frame moves with seasonal changes.

As you can imagine, it's harder to work on ceilings than walls. What is less well known is that it is much more difficult to install a smooth plaster ceiling than a textured finish with swirls or other designs brushed into it. The swirls look fancier, but they are actually easier and faster to apply. It is difficult to get a seamlessly smooth finish that doesn't show imperfections even to the untrained eye.

Gary McKay took charge of both the blueboarding and the plastering. He sub-subcontracted the blueboarding to Rick Wentworth. Blueboarders — wallboarders in general — have the worst reputation among the trades. I don't know whether their bad rap is deserved or not, but I didn't have any complaint with Rick and his crew. The four-by-twelve-foot sheets cover sur-

A "birdman" applies
plaster to blueboard
over the doorway from
the ell into the
all-seasons room. He
will work it over and
over with his trowel
until the surface is
smooth enough to
paint without first
sanding it.

faces in a hurry. The thing the boarders have to be most vigilant about is that in their fast-paced installation they don't accidentally cover up any electrical outlets or access to other mechanical systems. The sheets of blueboard are held in place with inch-and-a-quarter screws on inside walls. Since there is inch-thick insulation board covering the studs of the outside walls, the screws there are two and a quarter inches long so they can penetrate both blueboard and insulation board and get firmly seated in the studs. Corner bead is applied to all outside corners. The seams and inside corners are taped with white two-inch mesh tape that comes already glued on one side. There is no application of multiple coats of compound or "mud" on the seams.

Gary brought a good-sized crew in for the plastering: five plasterers and two assistants who mixed plaster in a plastic barrel with an electric mixer and brought it to the plasterers as needed. The very first thing they did clinched their reputation in my eyes — I knew I had the right people for my house. They covered all the floors with tar paper where they would be plastering. They hung heavy plastic sheeting over windows and window frames. Plastering is very messy. But when Gary's crew took up the protective coverings after a week of intense work, the house looked very neat. There was almost nothing to clean up afterward.

The plasterers hit all the ceilings and seams in the wall with a one-eighth-inch-thick base coat of plaster, leaving it a little rough to make it bond better with the final skim coat. Then ceilings and walls got a one-eighth-inch skim coat of gauging plaster (a special gypsum plaster mixed with lime putty for the finish skim coat); the final coat was troweled out four times with wooden-handled stainless steel trowels. They worked and reworked the plaster until it was so smooth that it shone with any light reflecting off it.

Gary told me that he deliberately set a slower pace than usual for his crew to ensure they did a first-rate job. Each plasterer did four "gauges" a day. A gauge is a batch placed on a guaging board from which the plasterer scoops it onto his trowel for application. The setting time for the plaster, once mixed, is about an hour. Gary did "the tough stuff" himself — for example, the large tapered skylight shaft in the main house just above the second-floor hallway. It was a very tricky area in his view, but not as difficult as the curving section of ceiling over the eyebrow-like window in my office. He "messed with" that small area for hours. When I recall that week of plastering, the image that sticks in my mind is of the plasterers walking about on their stilts. In a household attuned to the images of birdlife, they looked like cranes carefully picking their way along a shore or waterway. In my mind they became "birdmen."

George Hourihan, our painter, put a first brush to our house in November 1993. Through Tommy Silva's recommendation, George was hired to do the interior and exterior painting on one of our television projects in Lexington, Massachusetts. I met him on that project and liked his work. George also bid on the next television renovation we did in the Boston area, but his bid wasn't accepted. About the same time I said to Tommy, "Do you think George might have some time to paint my new house?" I hired him on a time-and-materials basis because I knew and trusted his working methods.

McKay's crew was in and out in a week on the plastering. George was in and out also as we needed him, but he spent a combined time of about three months on the interior painting, spread over a period of eight or nine months.

The story in painting is not in the sequence of surfaces that the painter touches. It is in his craftsmanship, his technique. George has had some good mentors who have taught him the techniques that distinguish a professional from a weekend do-it-yourselfer. One of his mentors taught him how to make his eyes follow the brush, how to caulk, how to dust down perfectly so that the surface to be painted is completely dust-free. On our house, George lightly hand-sanded every sharp edge on every window frame and every raised panel edge of every door to gently round off the edges; if you don't, he says, the paint gets very thin on the edge and shows as a line. He works from the most brightly lit surface back toward the less well lit. He paints crosswise to the path of natural light hitting a surface. Because of the different drying factors, he doesn't cut a baseboard until last when using flat paint, but he does the baseboard first with semi-gloss. All of these considerations enable him to get a beautiful, flat look to the surface, free of any brush marks.

George works with a roller in squares of about four feet to a side — it is about the largest surface he can handle and keep all of the edges wet. He doesn't roller the often recommended **W** shape and fill it in. He applies all of the paint in one direction, then cross-rollers it. It took George most of one day to paint the large skylight shaft above the second-floor hall in the main house. He studied carefully how light hit the shaft; after cutting in (carefully painting the edges or corners) the four sides of the shaft from an extension ladder, he rolled the surfaces standing in the hall and working with a long extension on his roller pole. In the same area, he rolled the tall curving wall along the stairwell sideways rather than up and down, working quickly so that the paint wouldn't dry too quickly and leave lapmarks where a section of fresh paint had visibly overlapped a section of partially dry paint.

Where George found any hardware already installed on doors and windows he was painting, he removed it before painting so that paint would not build up against the hardware. He painted both the tops and bottoms of all interior doors to keep them from swelling in humid weather. His routine when he had to sand somethng before painting it was to brush the surface with a dust brush, vacuum it with a mini-vacuum machine, and rub it with a soft cloth before applying any paint.

Just as Gary McKay covered everything with paper and plastic before he plastered, so George Hourihan carefully draped everything with dropcloths even though he was as neat in his work as the plasterers were messy. He knew everything and took every precaution. Well, not quite everything. At our house George learned one new fact for his craft. He had never painted a house with radiant under-floor heating. The heat was on one day as George was painting. He put latex paint in his steel roller pan and set the pan on the floor as he painted; very shortly he found his paint starting to thicken on him from the heat radiating from the floor through the steel pan. Now he knows to ask about a house's heating system before putting his pan on the floor.

It took him a week to paint the ceiling panels in the great room of the ell. He had to caulk the edges of each of the forty panels before painting it. At the peak of the roof he was working — alone — on a twenty-eight foot ladder set on the finish tiles; he admits to a little concern about whether the rubber grips on the bottom of the ladder would hold. You or I might think of taping the edges of those panels of wallboard to give us a pretty straight line as we painted them. George cut and feathered them in by hand — "That's my gift" — with a hand remarkable for its control.

For all of his meticulousness, George didn't avoid having a hand in one troublesome incident the week before we moved in — I suspect he'll remember the incident as long as he remembers painting our house. George was doing final painting that week in the bedrooms and bathrooms upstairs in the main house while the carpet men laid the Berber carpet. On Wednesday of that week George went into the walk-in closet off the master bath and reinstalled a plywood panel that he had painted; the panel covers a set of controls for the radiant heating on that floor. He experienced some difficulty setting the panel back into its frame but thought that maybe the paint had made it tight.

I was at the house into the evening. When I went through shutting off lights before leaving, I heard a funny dripping noise as I passed the powder room. Supposing it came from a fixture left running, I opened the door to find the floor being flooded with reddish solution that was dripping rapidly from the ceiling lighting fixture. I knew immediately that it was the antifreeze solution, which fills the tubing of the radiant floor heating, but where was it coming from?

I rushed upstairs to the walk-in closet, which is just above the powder room. More antifreeze was pouring out from the bottom of the panel covering the heating controls. I tried to remove the panel by lifting on the handle, but it wouldn't budge. I ran for a hammer and a scrap block of wood to tap on the wooden handle, but the only thing that happened was that the handle broke off. Not wanting to damage the panel beyond repair, I found a large screwdriver and drove it through the panel where the handle had been located, then lifted with all my might. The panel finally came free. When I looked inside I immediately saw the problem. The fluid was coming from a drain cap that was loose; the valve for the drain was open. I closed the valve and tightened the cap, stopping the flow of antifreeze. How did this happen? The panel had been installed previously and no antifreeze had leaked out.

After examining the problem a little closer I came to the conclusion that the valve had been turned slightly outward and the drain cap left loose when the plumbers had repaired some minor system leaks and purged the system a day or so earlier. They hadn't reinstalled the panel, so they would not have known that the valve had been turned enough to hit the back of the panel. George thought the panel was tight because he had painted it, but what was really happening was that it was hitting the valve handle, and when he pushed the panel down into place it opened the valve. If the drain cap had been tightened it wouldn't have leaked even if the valve were open, so I guess the problem was caused by both the plumbers and George. If I had not heard the dripping, the antifreeze would have flowed indefinitely, since the boiler has an automatic valve to refill the system when there is a drop in pressure. Who knows how many gallons would have leaked out before someone returned? A potentially messy situation if the antifreeze had flowed down the hallway under the wood floors was avoided. I thought about the irony of it as I drove home. A couple of the most meticulous craftsmen who worked on the house had unintentionally almost produced a disaster. That's one of the reasons I never get bored with construction work. All of the routines are interrupted regularly by the unexpected.

Finally, a Finished Floor

AS THE ROUGH-INS for the electrical and alarm systems were approaching completion in May of 1993, I began to install an additional layer of half-inch plywood on all of the floors of the house. It went down on top of the three-quarter-inch, tongue-in-groove plywood the framers had installed over the joists nearly a year previously to give us a working deck. This second layer of plywood would further strengthen the subflooring. I would lay the sheets perpendicular to the first layer. Wherever possible I made sure the long edge of each sheet fell at the center of the joists below. I applied a bead of construction adhesive on the subfloor in the area between the floor joists below and where the short ends of the half-inch plywood would fall. I then fastened the half-inch plywood down with one-and-a-quarter-inch screws at six inches on center along the short end of each sheet and at nine to twelve inches on center along the bead of construction adhesive between the joists. Wherever there was a joist below I used five-penny, ring-shanked nails, which I installed with my pneumatic nailer.

Where Laura and I had decided to install hardwood floors — in the living room, dining room, and library and the downstairs and upstairs halls of the main house and in my office in the back wing — I ripped two-by-fours in half to make strips called "sleepers" that were about one and a half by one and five-eighths inches thick. In the downstairs hall and in my office, because of the direction I wanted the finish flooring to run relative to the floor joists, I ran these sleepers every sixteen inches perpendicular to the joists and fastened them down with twelve-penny coated nails and two-and-a-half-inch screws wherever they crossed a joist. Again, for added stability in the subflooring structure, I put some construction adhesive between the sleepers and the plywood underneath. In the other hardwood floor locations, I ran the sleepers directly along the joists underneath. The sleepers were meant to be nailing strips to attach the finish flooring. Between the sleepers, the plumbers would clip down the tubing for radiant heating, and then another crew would pour gypsum concrete over the tubing up to the level of the top of the sleepers. Where any of the sleepers approached an outside wall, I left a little more than two inches of space so that there would be room for the one-inch rigid insulation to go to the subfloor and still be enough room for the radiant heating tubing to loop around the end of the sleeper into the next bay. Where the sleepers approached the interior walls I left a little more than an inch of space for the tubing.

Until Jock Gifford saw this system in another house and recommended it to me, I had been perplexed about how to have radiant under-floor heating with a hardwood floor. Either one would have to install a floating wood floor over the gypsum concrete containing the heating tubing or one would have to glue sections of parquet flooring directly onto the concrete, or so I thought. I didn't like either of those options. I immediately liked Jock's third option. Nailing the finish flooring to the clearly visible sleepers eliminated the possibility of puncturing the tubing accidentally, since no sleeper meant no nails.

The heating subcontractor's crew installed the tubing soon after I completed the plywood and sleepers phase. In rooms where the finish flooring was going to be carpet, tile, or vinyl, they laid out the tubing in predetermined patterns and spacing calculated during the design of the heating system. Basically they would start a designated area at the edge of the room and make their way to the center of the area at twice the specified spacing distance. Then they would reverse direction, centering the tubing between that which was already in place until they reached the edge of the room. Some rooms would have more than one area, since it was recommended that no single area exceed three hundred feet of tubing. On the floors where I had

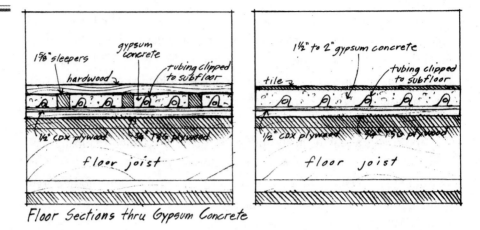

Floor Sections thru Gypsum Concrete

attached sleepers, obviously this type of pattern could not be employed; here the tubing ran up and back in each bay, looped around the end of a sleeper, and ran up and back the next bay. Probably the tubing isn't quite as densely employed in the patterns under the hardwood floors as under the other floors, but I wanted to be careful that I didn't put more heat than was really necessary under the hardwood floors — to minimize the drying and shrinking of the wood floors during seasons when the heat is on regularly.

Once the tubing is down, the sooner the gypsum concrete that becomes the thermal mass is poured over it, the better. Every day the tubed floors are walked over increases the chance that someone will stumble on a section of tubing and loosen the clips holding it to the plywood subfloor, or even accidentally puncture the tubing. The material we were going to use for the thermal mass was a nonstructural gypsum cement mixed with washed sand and potable water, forming a gypsum concrete formulated especially for installations over radiant heating. Different compressive strength levels can be obtained, depending on the ratio of prepared gypsum cement to sand as they are mixed on the site with water. I chose a higher than average strength level in order to protect the tubing embedded in it. Each square foot of one-and-one-quarter-inch thickness added about twelve pounds to the weight of the floor.

It took days of work to install the radiant tubing meticulously and several more days to install the batt and rigid insulation to all the outside walls, but

the gypsum concrete went down in a single day. I had prepped the house
with special care. All that the sales representative prescribed was a broom-
clean surface, but we vacuumed every floor as well. The installers arrived
early in the morning and snapped chalk lines around the perimeter of each
room at the desired finished height of the gypsum concrete. They also set
some grade marks in the middle of each floor area by driving nails into the
plywood subfloor; the top of the nail heads was the intended height of the
finished layer of gypsum concrete. Then they sprayed the plywood underlayment everywhere in the house with an adhesive to help bond the gypsum
concrete to the plywood.

The bags of gypsum cement are mixed with sand and water in a special
machine to a liquid that can be easily pumped through a hose — liquid
enough to be almost self-leveling after it's hosed onto a surface. The installers do, however, go around and make everything as smooth and level as possible with their long-handled screeds. Their plan was to pour the gypsum
concrete for the hardwood floors in one application, with the gypsum concrete coming up to the level of the sleepers.

On the other floor areas they planned a two-phase application. First, they
swept through the house from one end to the other applying enough gypsum
concrete to cover the tubing. By the time they got to the far end, the first
application where they began had set enough that they could return to the
starting place, checking along the way to make sure there were no "floaters,"
or sections of tubing that had come unclipped and floated to the surface of
the liquid gypsum concrete. Then they sprayed on another thin layer of adhesive and began hosing on the second and final layer of gypsum concrete.

There was nothing wrong with the plan. But just as they were well along
on the first pour — both floors of the main house were poured and they were
halfway across the great room of the ell — the mixing machine broke. Sometimes a piece of rock in the sand jams the mixing blade, damaging it or the
shaft that drives it. That's what happened in this case. The machine was in-
operable until the crew could get to a welder. They had come from a great
enough distance that it was impractical to return to their own shop for another mixer or to make repairs. Fortunately, the town garage is located close
to one end of our subdivision. We found a friendly welder there willing to

Gypsum concrete, mixed outside the house and pumped in through a thick hose, begins to cover the radiant heating tubing. While it is still malleable, a technician (opposite page) wearing home-made footgear smooths it with a screed.

perform the quick weld needed. It took about two hours to diagnose the problem, find the welder, and get operational again. The delay complicated the day but didn't prevent the crew from finishing the installation of all of the gypsum concrete.

Ideal curing conditions for gypsum concrete are opposite to those of ordinary concrete. The more slowly concrete dries and hardens, the stronger the concrete becomes. Gypsum concrete needs to set and dry quickly to reach its greatest strength. At the end of the day we left all of the windows in the house open to let moisture from the concrete be drawn outside. I had hoped to have the zoned air-conditioning system running by the time the installers arrived, but the gypsum concrete had to dry only by natural ventilation for a day or two. Then, with the air-conditioning in service, I closed all of the windows and ran it full-time for a few days to dry the concrete as fast as I could.

When Joe Ferrante came to the house with a crew to install the tile for the floors in three of the bathrooms, the great room of the ell, the all-seasons room, and the hallway and laundry room of the office wing, he advised putting down a coat of special material between the gypsum concrete and the thin-set tile cement that would be a bed for the tile. This extra layer constituted an elastomeric (or flexible) membrane that eliminates tile cracking caused by structural movement and acts as a waterproofing membrane for the gypsum concrete. Since we had selected a sixteen-inch-square tile for these areas, we were especially concerned about potential for cracking.

There are several Ferrantes involved in the family business, but I dealt with Joe, whom I first met when he tiled rooms in houses we were videotaping for *This Old House* programs. Laura undertook the shopping for any tile, vinyl, and carpeting we needed. In the beginning, she shopped for tile at a showroom in New Hampshire, but while the store had a very large selection of tiles to choose from, it also had a policy of not offering samples to customers to take home to look at in place, and it didn't permit returns of any tile once taken from the store. We still have enough tile for our master bathroom floor sitting in boxes in the garage. Laura and I both thought the tile looked perfect — until we placed several pieces of the tile on the floor to check it out before Joe set it in place. The floor tile looked too gray. Unacceptably gray. So beware of tile suppliers with a no-samples-no-returns policy. Eventually Laura found a white tile that we installed in the master bathroom.

At first, Laura looked for terra-cotta tiles for the great room floor. Their warm reddish brown color would surely work well with the several warm wood tones in the timber-framed space. However, we were concerned about the maintenance of terra-cotta and its potential for absorbing stains. Laura found a few terra-cotta lookalikes, but they didn't have the diversity of surface and coloring that is one of the virtues of true terra-cotta; they looked as though they had all been pressed from the same mold. As eager as we were to get the tile down in the great room and the all-seasons room, we still waited until the kitchen and family room cabinetry had been installed. That way we could look at tile samples placed next to the wood of the cabinets and under different lighting conditions. To take a chance selecting and installing over twelve hundred square feet of tile with just a small sample of the stained wood for our cabinetry constituted a risk much too big to take. As it turned out, the wait allowed us to visit a tile supplier Joe recommended who would allow us to bring as many samples as we wanted to the house for a comparative look. It pricks my Yankee conscience to think that I have a bathroom floor's worth of tile sitting unused in the garage; a great room supply would have sent me into a tailspin. Eventually we chose an Italian glazed ceramic tile. Like terra-cotta, it has more of a matte finish look than a hard, shiny look. Like terra-cotta, it has individuality of color from one tile to the next. The colors are very soft — grays and very pale dusty pinks in each square. In a space as large as the great room, the soft cloudlike patterns in the tiles seem to pull the space together better than a tile of a single, uniform color.

The Ferrante crew troweled on the special coating, then let it set and dry for a day before laying any tile. There was nothing tricky about the installation of the tile. They set their guidelines and then troweled thin-set coating onto areas large enough for a few tiles at a time. As tile went down in the great room, we could immediately see that the scale of the large tiles fit the oversize scale of the room, and it appealed to us to have the continuity of a single kind of tile wending its way from the all-seasons room through the great room, on into the hallway in the back wing, and finally through the laundry room.

It is important to avoid walking on fresh tile for a day or two while it sets, so there was a brief delay before the Ferrantes returned to apply grout between the tiles. The grout also needed a day or two to set so that the flexing of the tiles as someone walked across the floor would not break the connection between grouting and tile before they were thoroughly bonded to each other.

I conducted the search for wood flooring myself, getting samples that Laura and I could place in various rooms and evaluate. From the very beginning, I was interested in having some recycled longleaf Southern Yellow pine on the floors of our house. It is a beautiful wood that I had seen installed in several renovation projects and I had even used some for furniture projects. However, I didn't think it was right for every one of the rooms we were flooring in wood. Another wood I favored was white oak. I really like its look. One of my dreams early in the planning of our house had been to have the timber frame of the ell executed in white oak. The lack of supply of good-quality heavy white oak timbers quickly shattered that dream. But there was no shortage of high-quality white oak flooring.

Several discussions Laura and I had about wood floors yielded a consensus. The formal living room, dining room, and library would be floored in white oak, but rather than using the conventional two-and-a-quarter-inch-width boards, we chose a pattern of alternating boards of three, four, and five inches, quarter-sawn. "Quarter-sawn" refers to the angle of the blade to the grain as the boards are milled; it gives a very beautiful look to the grain and the floorboards are less likely to warp or "cup." I got Laura to agree to our using the Southern Yellow pine in the two halls, upstairs and downstairs, in the main house, and on the treads for the stairs joining the two halls.

Jeff Hosking of Walpole was my choice to install our wood floors and to help me locate the materials I needed. He depends heavily on an excellent supplier in Connecticut named Charlie Wilson. Charlie obtained some Southern Yellow pine samples from a Florida dealer I knew of, who salvages sinker logs from riverbeds. Sinkers are logs that were lashed together with many others to float down a river from forest to mill but then broke away in passage through shoals and remained trapped underwater. Some logs have been trapped for more than one hundred years, but because of the high pitch content of the wood and because they were totally submerged free from exposure to the atmosphere, they remain in excellent condition. However, the larger portion of old-growth Southern Yellow pine comes from beams salvaged from obsolete mills and warehouses that have been demolished. As this wood has grown popular in the past several years, the supply has diminished and the price has risen. I rejected the Florida samples, not because of quality but because the price was out of sight.

After a lot of research and phone calls, Jeff found a dealer in South Carolina who sent me fully acceptable samples and prices. Jeff placed an order to be delivered already milled to his shop. The six-inch flooring arrived at Jeff's shop in mid-March 1994. I borrowed a pickup and moved the flooring from Jeff's shop to the house, and then I "stickered" the pine — that is, I stacked it in the house with thin strips of wood between the boards to allow air to circulate fully around them. Several weeks of getting acclimated to our house before it was installed would be good for the wood, I thought. It took a little longer to get the thicker five-quarter-by-six pine boards. I had those delivered to a workshop, where I could edge-glue pairs of boards together and then mill the blanks into stair treads for the main staircase before I brought them to the house.

The white oak came from Charlie Wilson's business in Connecticut. It was delivered February 27, and the following weekend my son Bobby stickered the correct amount of flooring in the dining room, living room, and library near the wall opposite from where I anticipated Jeff would start installing the flooring. Before Jeff started nailing down any flooring, I checked the sub-flooring. I noticed that in many places the gypsum concrete was slightly higher than the wood sleepers to which the flooring was to be nailed, so that the flooring would actually rest on the concrete in all those places rather than on the sleepers. It wasn't that the installers had erred and poured the concrete too high. In the several months during which the concrete and sleepers had sat there waiting for the finish floor — with some heat on in the house during cold weather — the wood of the sleepers that I had ripped from two-by-fours had dried and shrunk.

I decided to rip quarter-inch plywood into inch-and-a-half-wide strips and nail these strips to the tops of the sleepers to bring them above the level of the gypsum concrete. I think in hindsight it was a good thing to do for a

Jeff Hosking installs quarter-sawn oak flooring in the dining room; I had had the wood stacked in the house for several weeks.

couple of reasons. The oak and pine flooring now doesn't touch the concrete anywhere. There is a slight air pocket underneath the floor. I think that because my extra step eliminates any contact between the flooring and the concrete, the wood will absorb less of any moisture that might be present in the concrete, helping to minimize movement of the flooring during climatic changes.

April 4 was the day Jeff started installing flooring. He installed the oak floors first. The boards were very well milled. The only hangup was that we needed to have the hearthstones for the living room and library in place before he brought the flooring up to the fireplaces. The stone and marble supplier was very slow about cutting and polishing the hearthstones for us, despite periodic urgings, and Jeff was within two feet of the living room fireplace when its hearthstone arrived in the nick of time. Almost as soon as Lenny Belliveau set the hearthstones on a bed of mortar, Jeff installed a five-inch border around the hearth.

Before Jeff installed any of the Southern Yellow pine I decided to have my son Bobby treat the underside of the flooring with a couple of coats of shellac. We were using pine flooring that was six inches wide, and I felt that the wider pine would react more dramatically than the narrower and harder quarter-sawn oak to any moisture present in the concrete. The quick-drying shellac would provide a moisture barrier for the underside of the flooring. It took Bobby only a few hours to get the job done.

The Southern Yellow pine gave Jeff more challenge than the oak. There were some varying widths among pieces that were supposed to be of equal width, making me regret that there hadn't been a way to have the pine milled by Charlie Wilson. But, at a little slower pace, down it went in the lower hall and part of the upper hall. Jeff couldn't finish the upstairs hall near the stairwell because I hadn't had time to do the carpentry on the stairs.

We didn't stain any of the floors, preferring to look at their natural colors as brought out by a clear finish. It was my decision to use a solvent-based finish on the wood floors rather than a water-based finish because I believe it brings out the color of the pine better. If we had already moved the birds into the house, I wouldn't have done that. Birds are more sensitive to toxic substances in the air than humans are. I think of the old-style mining days when miners took birds underground because the birds reacted swiftly to toxic gases in the tunnels before humans were necessarily aware of the danger. Our birds were not exposed to the fumes from the three coats of polyurethane Jeff brushed on the pine and oak floors.

When the floors were down and finished, anyone could see the difference in color between the two woods. The white oak is a medium-dark sandy color, a very rich tannish tone. The longleaf Southern Yellow pine has a lot of reds and yellows in it, a more dramatic color than the oak. The two woods meet at the openings of the hallway into the living room, dining room, and library. I don't mind the difference in color at all. Both woods appeal to me. Laura likes the oak more than the pine, but I think the pine is growing on her, and I think when I have finished installing the finish wood of the treads of the stairway so that the halls and stairway are one continuous runner of pine, she will be captured by its beauty. (Maybe it's not the pine but the unfinished state of the stairs that bothers her . . .)

Eventually, my office will have a wood floor, too, but not until some time after we take up residence. Since my shop won't have been built when Laura, Lindsey, and I move into the new house, I will have to use my new office space as both home study and workshop for a while. I will keep some power tools there and move them about the house as I work on finish details, here some shelves, there some moldings, over there some wainscotting. If I follow through on my present intention, I will one day order cherry wood for the office floor.

By June of 1994, all the floors were accounted for but the bedrooms in the main house, and the bird room in the office wing. The bedrooms were to be carpeted. The bird room was going to get a covering of practical vinyl, easy to maintain. Laura shopped for carpet and vinyl at the same flooring outlet in a nearby town. Both these coverings were installed the week before we moved in. Installers came from the business where we bought the coverings, but of course carpet men are one breed and vinyl men another, so there were two different crews.

There was plenty for my father and me to do before the carpet was laid. We hung the doors, applied the finish trim to the windows and doors, and installed the baseboards in all four bedrooms, and I had George Hourihan finish the painting in these rooms. Everywhere that carpet or vinyl was to be installed, I checked the gypsum concrete subflooring carefully, picking up any blobs of plaster that might have gotten under the tar paper that the plasterers used for protecting the floors, scraping down any high spots in the gypsum concrete, and then vacuuming the surface carefully.

When there is a lot of traffic on gypsum concrete, its surface gives off dust particles. I talked to the gypsum concrete installer, and he gave me a latex sealer to apply before the installation of carpet or vinyl. It came as a liquid that I diluted with water and sprayed on, two coats, with a pressurized tank-type unit made for garden spraying. The sealed concrete won't wear down as fast if the surface is prevented from sloughing off. Also, the sealed surface was a better surface than untreated gypsum concrete for bonding to adhesives. Obviously this fact is relevant to the bird room, where the vinyl men would lay the floor covering over a thin bed of adhesive. But it had consequences for the carpeted rooms as well. The wall-to-wall carpeting in the bedrooms is held in place by impaling the carpet on tack strips attached to the subflooring around the perimeters of the rooms. These strips replace the old way of attaching a carpet to the floor by driving tacks from the topside of the carpet into the floor. Usually the tack strips are nailed down. We could have used nails long enough to extend through the gypsum concrete and into the plywood, but there was a risk that nails might puncture some of the radiant tubing embedded in the concrete. In light of this risk, I asked the carpet installers to fasten the tack strips down with construction adhesive.

The medium-gray patterned vinyl went down fast. The installers checked the gypsum concrete surface carefully for any gouges and filled them; otherwise, the relatively soft vinyl would over time telescope down into any indentations. I was pleased to see how perfectly parallel to the walls the pattern runs. Sometimes installers will get off a little — usually from mismeasuring or from choosing the wrong wall for orientation when the room isn't completely square. Our house was new and the walls were square, but even in new construction I've seen vinyl and tile installations go astray. Except for one seam in the floor where the installers used a seam sealer that needed to set overnight, the vinyl floor could be safely walked on immediately after its installation.

The day the vinyl went down, the carpet installers arrived to glue the tack strips to the gypsum concrete so that the glue would dry overnight and they could start installing the carpet the next day. Laura and I both preferred a good wool Berber carpeting. We chose a natural fiber, wool, over synthetics because we had heard and read about out-gassing of chemicals from synthetic materials. Again, our concern was as much for the health of the parrots as for our own. There was a considerable range of price in the carpet samples we saw, correlated closely with the thickness of the carpet (the thicker, the pricier). Fortunately for our now besieged bank account, a middle-range carpet was the best choice. I didn't want to put too much carpet on the floor, because it would act as an insulator between the radiant heat rising from underneath the carpet to heat the room. Since we had installed radiant heating, we couldn't use foam padding under the carpet, so we elected instead to use the familiar old jute pads. Laura chose to use the same creamy beige Berber carpeting in all of the bedrooms, and I like the effect. Identical carpeting seems to pull the bedrooms together visually, just as the tile down-stairs visually unites the great room and the all-seasons room, and the Southern Yellow pine flooring makes two halls and a stairway in between a continuous path.

It remains to be experienced whether the radiant heating will heat the bedrooms easily in bitterly cold weather. I believe it will. But if there is an especially cold night, I have a fallback plan in the master bedroom. I'll light a small fire in the fireplace there, and sleep in comfort.

The Inspector Calls

THERE WERE FIVE of them poking around the site between the time we began to clear the site and the time we moved into our new house — inspectors, I mean. The number includes two different persons from the Conservation Commission because of a staff turnover; but there could well have been five in any case because the building inspector doubles as the electrical inspector in our town.

Bob Koning, the building inspector for the past fourteen years, is almost overqualified to be the electrical inspector. He started an electrical contracting business in Boston more than thirty years ago, which some of his children still run with his part-time assistance. Being building inspector in this town where he has resided for forty-one years is, in his words, "a hobby that got out of hand." Somehow he also finds time to be fire chief of the all-volunteer fire department. Given his schedule, I think residential fires around here may be permitted only by appointment.

Tuesday and Thursday mornings Bob makes himself available for consultation in his office in the town hall, and on those afternoons he usually makes his rounds for on-site inspections. He doesn't have to drive very far for any of his visits. The town is only three miles square and includes an eight-hundred-acre state park, the county's only cranberry bog, lots of wetland, and housing for only about 4,200 people and their pets. In the 1950s the citizenry endorsed two-acre–minimum plot zoning for any additional houses, the only town in the area with more than one-acre zoning; half-acre zoning is most common in the surrounding larger towns. There is no commercial zoning in our town except for the half-dozen properties near the village center, which were already the sites of little businesses in 1952.

In a typical year Bob inspects between twenty and thirty new houses, among which ours is of average size — if anything, on the smallish side. There was one year in the late 1980s when there were seventy-five houses under construction, and Bob was going crazy. Then the building recession struck, and the next year there were only five or six. "It all averages out." Land has become so valuable here that people occasionally buy properties, tear down houses built, say, in the 1960s, and replace them with larger houses for the 1990s.

We started digging the foundation for our house before we actually got the general building permit. But before we could pour any footings for the foundation, Bob came to check the soil conditions under the footings and, peculiar to our site, the pins we had installed in the exposed granite ledge in the northeast corner of the basement to ensure that the foundation wouldn't slip where it sat on the ledge. Once he signed off on the bearing quality of the soil and the stability of the pins, we could have the footings and foundation walls poured.

Then Bob didn't return for a while. Some towns require an inspection of the finished foundation before the builder can backfill it, and if there is damp-proofing applied on the outside of the foundation, it must be inspected, too. The number of inspections is to a degree a matter of judgment on the part of the inspector. I suppose his approach to inspection of each house is influenced by the confidence he feels about the first stages: the degree of detail in the architectural plans, the experience of the builder, and other such factors. When I asked him one day what is the biggest problem he encounters as an inspector, he said, "Nonqualified people trying to build a house. They begin with a lousy set of plans. They award the job automatically to the low bidder without any thought to his ability and record. Rather predictably, they run into problems. In the 1980s anyone with a saw in his hand thought he was a builder."

I don't know why Bob elected not to inspect our finished foundation. He told me about once inspecting a foundation where the excavator hadn't bothered to make the floor of the hole quite level. The foundation subcontractor came in to pour the footings and knew he had to make those footings level — which he did. Then he poured the foundation walls a short time later. When Bob checked the finished foundation he saw that the footings at one end of the foundation conformed to the minimum requirements in the state building code: they were ten inches thick. But at the other end of the foundation the footings got as thin as three inches. It isn't fun to tell a builder that a large section of a finished foundation has to be torn out and rebuilt with footings that conform to code.

When Bob came back for his second inspection — the framing inspection — the house was ready for insulation. The walls were up, the roof was on, the windows were installed. The house, in short, was weather-tight because it is foolish to install insulation before reaching that point. In the meantime, however, the plumbing inspector had called.

James Sullivan, the plumbing inspector, works under the general supervision of Bob Koning. Our plumbing contractor, White Rock, took responsibility for getting the plumbing permit required before any rough plumbing could be installed; and White Rock also managed the process of scheduling inspections and getting the inspector to check off the relevant boxes on the building permit mounted in a window at the front of the house. Sullivan came a couple of times. On his first visit, he checked the hot and cold water supplies to make sure they were piped and sized correctly.

Sullivan's inspection of the waste removal system involved a rigorous test that required the collaboration of the plumbers. Ordinarily there is no standing water in the waste system, except for the small amount of water retained by design in the fixture traps to prevent sewer gas from backing up through open fixtures into the house. As waste is discharged from bathrooms, kitchen, or laundry room, it flows by gravity immediately through the system and out of the house toward the septic tank. Sullivan had the plumbers block off the main waste pipe where it exits the house in the basement. Then the plumbers took hoses up on the roof, where the open vent pipes marked the highest point of the waste removal system. They hosed water into the vent pipes until the entire system was full. This procedure creates enough pressure in the lines to blow out any inadequately soldered joints. White Rock had done their work well. There were hundreds of connections where they could have made a small error, but no leaks were discovered in the lines while standing water exerted heavy pressure on the entire system.

I find it odd, by the way, that there is no inspection of the heating system mandated in the state building code. There is so much that can go wrong with improperly installed boilers; heating systems generally require more maintenance and repair than any other mechanical system in the house. The building code, however, seems to be silent on the subject.

When James Sullivan came back for what I intended as the final plumbing inspection before the final general building inspection to qualify us for an occupancy permit, I was worried that he might object to the fact that I was deliberately leaving two of the four and a half bathrooms (the bathroom off the guest bedroom and the bathroom next to my office in the back wing) in the house unfinished — untiled, pipes for fixtures installed but capped. I believe the Board of Health regulations and the building code require only that there be one functioning bathroom in the house. But you never know what will make an inspector unhappy. Sullivan, however, didn't question the unfinished bathrooms.

Just as White Rock had gone about getting the plumbing permit for the house and then scheduling the plumbing inspections, Bob Russell, our electrical contractor, attended to the electrical permits and most of the inspections. It's possible to get one permit to cover the whole electrical installation, but in that case the builder has to know how many electrical fixtures — outlets, lights, switches, and so on — are going to be installed. In the chronology of building our house, I found that I wanted to get the power box in the house before we had completed the electrical plan. So we reverted to two stages of permits, getting authorization only to bring the electrical service as far as the utility room under the first permit.

Bob Koning, wearing his electrical inspector's hat, came once to check the work done under the first permit. He first checked the conduit through which the cable would come underground from the nearest transformer to the meter on the west side of the garage and then under the garage floor to the electrical panel board in the utility room. When he saw that the plastic pipe was trenched deep enough outside (minimum two feet below surface), and surrounded by a few inches of sand between pipe and the nearest soil, he signed off on backfilling the trench. At the same time, he inspected the panel-board installation as a prerequisite to getting the power turned on. This stage tried my patience with the amount of coordination exacted by the various parties. The local utility has one group that just installs the meter to the house and another group that makes the physical connection between our cable and the transformer out on the access drive. Neither of these groups can do anything until the electrical inspector authorizes the connection; then they might do something, depending on their already full appointment books.

The second electrical permit Bob Russell got was for the rough electrical wiring. When he had run wires to all of the fixture locations and installed boxes at those locations, but hadn't installed any switches or plugs or lighting or other fixtures — and before the system was "alive" beyond the panel in the utility room — Bob Koning came back for another inspection.

The framing inspection, Koning's second inspection as building inspector, couldn't logically occur before the rough plumbing and rough electrical work had been completed. That's because the electricians drill holes through the wood frame to run their wires. The plumbers make even bigger holes than the electricians. The building inspector wants to make sure than none of these holes drilled, or notches cut, in studs and joists has compromised the integrity of the frame. He looks for other flaws in the frame, too. Have the framing materials been sized properly? Is bracing adequate everywhere? Have proper clearances been maintained between fireplace boxes or flues and any surrounding wood or other combustible materials? Our structure passed the test pretty cleanly. The only items specified to be added — nothing had to be corrected — were the provision of a ladder into the attic space above the bird room in the office wing, and the mortaring in of the support beams in the basement where they went into the foundation, so they couldn't move sideways or tilt.

I would have been surprised if Bob Koning's inspection had revealed anything defective, because I believed the various subcontractors and I were familiar enough with the building code not to violate any of its major provisions. My intention was to build a house of far greater quality than the building code assumes or requires. The code is more of a minimum standard for certain specific aspects of construction. It is possible to build a house of, in my opinion, poor quality and still get it past inspections because it will, despite its flaws, not violate code.

When I was active as a general contractor myself, the framing inspection was usually the final visit by the building inspector until he came for the final inspection before issuing an occupancy permit. But a few years ago, the state added a step of inspecting the insulation after its installation, so Bob Koning had to return for another walk-through after the rigid insulation was installed on the inside of the frame of exterior walls and in the attics.

Between Bob Koning's insulation inspection and his final inspection before occupancy, the Conservation Commission called again in the person of Christine Chisholm, a new staffperson who had replaced Pat Loring. When she was in our neighborhood one day on another errand, Christine stopped by to check the silt barrier we had erected between the construction site and the wetland to the west of it. She telephoned later to say she was concerned that the hay bales in the barrier were deteriorating — not surprising, since they had been exposed to weather conditions for more than two years. I didn't want to take any chances of running afoul of the commission as we neared occupancy, so I made an appointment for Christine to return when I could be there to look at the barrier with her. She was a very reasonable person to deal with.

Work on the house lagged in the winter months of 1994 because I was in Hawaii much of January, February, and March televising programs for *This Old House*. When I got back from Hawaii at the end of March, I intensified my own efforts as well as the work of some of the subcontractors to get the house ready for occupancy. I almost decided to try to get an occupancy permit in May, but several of the light fixtures hadn't been acquired or installed yet. Bob Koning would be wearing both his hats when he came — doing the final electrical inspection and the pre-occupancy building inspection in the same visit. It was going to take from two to three weeks to obtain the lighting fixtures and get them installed. Bob wouldn't sign off on the electrical work as long as there were any open boxes. We had, moreover, just ordered carpet

that Laura had scouted for the bedrooms; installation of the carpet was a month away, which meant that occupancy was at least a month away. I decided to postpone the occupancy inspection about thirty days. Better not to risk doing one of the things I believe irritates inspectors most: calling them in before you're ready.

Winter 1994. The house stood quietly in a bed of snow while I videotaped television shows in Hawaii. Work would resume in early spring.

When I called Bob a couple of weeks ahead of my target date to schedule the inspection, I asked him what he would be most concerned about as he toured the house. He reminded me to get written sign-offs from the Board of Health and the Conservation Commission. He listed a few obvious things, such as being sure the fire and smoke alarms were operational. He mentioned one requirement that was amusing, namely, that there be doorknobs on at least two exterior doors. I hadn't, in fact, yet installed all of the hardware on interior doors, but I couldn't imagine anyone requesting occupancy of a house lacking a full complement of exterior doorknobs. Inspectors must see some very curious situations.

What worried me most was that none of the three stairways in the house would be completed before we moved in. The finish carpentry of the stairs was going to be a time-consuming task, and I didn't want to compromise the final outcome by slapping on hastily built finish woodwork. For some time we are going to be walking up and down plywood steps. Bob said that plywood stairs were acceptable so long as each open stairway had a handrail and that there were no open risers between the treads. When we spoke, *all* of the stairs had open risers and the main stairway in the main house, lacking a handrail, looked like a liability insurance agent's nightmare.

The day before the occupancy inspection I ripped up plywood to make temporary risers. I jerry-built a main house solid handrail out of plywood sheets and two-by-four posts. (My two dictionaries refer to "jerry-built" variously as hastily, cheaply, and poorly or flimsily built. Well, my temporary railing was built hastily and cheaply but not poorly or flimsily. It's probably stronger than the finished handrail and balusters will be.)

No single aspect of my preparation for the inspector's visit took more time than finishing the ceiling of the utility room. The state building code requires the use of heavy fire-code drywall there to inhibit the spread of any fires originating in the room. We installed that ceiling with the proper material early in the construction process, but then the plumber removed a couple of the panels when he ran the pipes for the heating system. Now, in hot weather, I had to spend two days cutting and piecing drywall around the pipes and then taping around all the places where the pipes went through the drywall. It was painstaking work in a pretty inaccessible place. I guess I accomplished it satisfactorily because Bob Koning scarcely glanced at it during his inspection.

The plumbing inspector was supposed to return to check the control valve on the master bathroom hand-held shower attachment on the 23rd of June; then Bob Koning was to visit us on Tuesday, June 24. I was nervously contemplating the consequences of his failure to appear, when James Sullivan finally arrived during the afternoon. His inspection took five minutes — in and out. Now we were ready, except for a few lighting fixtures, which our electrician, Bob Russell, had promised to install on the morning of the 24th.

We must have been first on Bob Koning's list that day. He arrived about half past one o'clock. I was a little nervous, but he was very relaxed and friendly. There were a few details that he could have picked up on if he had been in a mood to be persnickety. For example, there were no dampers on the four fireplaces. I had ordered them custom-made, but they hadn't arrived. Their absence was not conspicuous because the opening from the firebox to the flue in a Rumford fireplace is comparatively small. If Bob noticed the absence of the dampers he elected not to make an issue of it. He was extremely interested in the radiant heating system, which wasn't hidden yet in the bedrooms because the carpet hadn't been laid. I thought he would have seen a fair amount of radiant heating in new construction of large houses in our town, but he hadn't. He liked the kitchen ("one of the nicest I've seen," he commented) and the timber-frame construction in the ell. It was, he assured me, a very well built house.

I followed along as Bob walked methodically — not hurrying but not dawdling either — through the entire house. Then we engaged in brief small talk — and it was over. Quick and easy. I felt a sense of relief, and yet it was almost disappointing that a ritual that symbolized the conclusion of such a long and ultimately fulfilling construction should be so routine.

Perhaps the routineness of the inspection made me hesitant to believe that we were authorized by the state to take occupancy until I actually held the certificate of occupancy in my hand. So the next morning I dropped Lindsey at school for one of the last days of her academic year, returned to the new house to work for a couple of hours, and then headed down to the town hall to pick up the certificate. It wasn't ready. Bob Koning's secretary prepares it, but Bob has to sign it and he hadn't come in yet. It wasn't one of his regular days to have office hours. Laura picked up the certificate the next day. What an insignificant slip of paper to represent a dream come true. But it meant something. It meant we could move in anytime — anytime after we got the carpet installed, that is.

*In place of a corner-
stone outside, we have
the handsomely in-
cised date on the truss
joist nearest the fire-
place in the great
room of the ell. 1992,
the year the trusses
were made and in-
stalled, was midway
between our beginning
to plan the house and
our moving in.*

Moving In

IT HAD BEEN more than four years since Laura and I first talked about moving in 1990. No one could blame me if I felt eager to move in and settle down. But there were two matters that stood in the way.

First, we had to assess what it would be like to move into a house with much finish work undone. Could Laura and Lindsey tolerate the noise and mess while mantels, bookshelves, moldings, chair rails or wainscotting, and other finish items were cut, installed, and stained or painted? These were all items I wanted to build myself as time permitted.

With my various commitments, "as time permitted" would certainly extend over a period of several months or more. Fortunately, the ell with its open kitchen, family dining area, and family room was complete, except for the installation of the tile/backsplash in the kitchen, one window screen, and the screen doors. The family bedrooms and bathrooms in the main house were complete, except for rehanging the bedroom doors, which I had removed earlier in the week for the convenience of the carpet men. Everyday life in these rooms would be spared the invasion of table saws and nail guns. If I worked on finish carpentry in one of the other rooms, it needn't be intolerable — or so I hoped. I'd take tools and materials in, close the door, and work to my heart's content.

One temporary structure did promise to be a little bothersome. The stairway from the first to the second floor of the main house had temporary ply-wood treads and risers, and a temporary railing framed with pieces of two-by-fours and filled in with panels of plywood. I'd have to count on Laura and Lindsey to be patient with the disruption when I got a chance to cut, install, and finish the recycled longleaf Southern Yellow pine treads, the painted soft pine risers, and the painted birch balusters and stained birch railing.

A second consideration was that we were moving into a larger house, for many rooms of which Laura and I planned to acquire new furniture and furnishings. But there had been very little time for furniture shopping. Selecting fixtures and floor coverings had exhausted most of our shopping time. There were things we could move from the old house and use until we had time to do more shopping. Of necessity, though, the new house was going to be somewhat sparsely furnished for a few months. Laura brought dishes, utensils, and supplies for the kitchen in carloads over the course of a few weeks, so we knew there was one well-equipped area even before moving day.

Our moving strategy was determined not so much by the family's needs as by the needs of our "extended family" of parrots. The parrots had to be moved in one swift, well-orchestrated event. When they were transported to their new home in our new house, we had to move in permanently to care for them.

I proposed Friday, July 1, because it was the only time during the summer that I saw four open days in a row in my schedule. We weren't videotaping any programs on Friday because the crew were leaving early for the holiday weekend. Monday was a national holiday. There was one reason why July 1 might not be ideal. Many people move on the first day of the month, so it's hard to get equipment. On Monday before the Friday move — talk about waiting until the last minute! — I started calling truck rental agencies. The first place I called had no equipment to rent for Friday. The second place had one unreserved truck. The agent wasn't sure I could have it for both Friday and Saturday, but reserved it for Friday in my name.

Between Monday and Friday there was much to be done to get the house ready. George Hourihan came early in the week to put a final coat of paint on the trim and touch up the walls and ceiling of the bird room. Then he spent

Peter awaits the move from the security of his own private car.

a good part of the week dodging the carpet installers while he put final touches of paint on bedrooms, the master bath, and Lindsey's bathroom.

On Wednesday, a medium-dark sheet vinyl with several shades of gray in its geometric pattern was laid in the bird room; with the dove-white walls, it gave the room a background against which the birds' feathers would look brilliant. Carpet installers came the same day to glue tackless strips, which hold the carpet in place, onto the gypsum concrete underlayment in the bedrooms. The glue dries overnight so that carpet can be installed the next day.

Thursday, the installers returned to lay carpet in all four bedrooms. I was at the house on Thursday until after midnight doing last-minute cleaning.

Shortly after eight o'clock on Friday morning, Laura drove me over to the truck rental agency to pick up the reserved truck. Before we left, however, Lindsey came out to say that the rental agency was trying to reach me. I was immediately apprehensive that the agency was about to renege on the reservation. When equipment is in great demand, all it takes is one renter failing to return a truck on schedule to set off a disastrous chain reaction. Laura took the call and learned that there had been an equipment delay but that by the time we reached the agency a truck should be available.

The rental truck was meant to carry the parrots' cages but not the parrots. Laura knew the birds would be thoroughly spooked if they were transported in the darkness of the storage compartment of a truck. All of the birds have been transported from time to time for such purposes as visits to the avian vet Laura uses, but only singly or in small groups in a well-lit car. Laura had engineered the moving ceremony to be just like a trip to the vet. Several of her friends came in their cars to help with the transfer. Each vehicle would have a parrot or two at most in carrying cases, so that the driver or a passenger could calm the birds en route.

Because Laura had planned the move meticulously, the loading went smoothly and quickly. Yet it still yielded the tensest moment of the day. I looked at the truck I had rented. The cages had been loaded faster than I anticipated. The truck still had most of its storage space available. There were several able-bodied persons on the premises. What an opportunity! The helpers and I could load some furniture in the rest of the truck compartment and make this trip really efficient. How shall I put this diplomatically? Laura was not pleased with my proposal to delay departure while we filled the truck. Rather quickly I was in the truck ready to pull out at the head of the caravan.

I don't know whose "law" applies here, but I'm certain it's true that in any operation planned as thoroughly and rationally as Laura planned this move, something is sure to go awry. Our single caravan broke, for various reasons, into three different sections — two of which went badly astray before making their way to their target. Yet all arrived, though not before Laura had some anxious moments.

The unloading scene was very busy but orderly. As soon as their cages were in place, the parrots were brought in from the vehicles in their carrying containers. Lindsey began to get fresh water and food into the cages. Richard Howard was moving around taking photographs of the entire operation. I knew that Laura was beginning to relax when she called out to me — maybe half in jest? — "I think the bird room is too small, Norm. Maybe we'll have to put some of them in your office."

*The front door is
officially open.*

It wasn't yet one o'clock in the afternoon when the last of Laura's friends left. Then I realized how much better Laura's slight overplanning had been than my substantial underplanning. There wasn't a piece of furniture for the humans in the house except one new box spring and mattress I had quietly loaded into a van in our caravan. I still had the rented truck but no helpers.

Lindsey agreed to help me. We drove the rented truck back to the old house. I selected a few heavy items that we needed for basic comfort — a sofa, a dresser, enough beds for the three of us — and Lindsey and I got them out of the house into the truck. I added other items, such as a cooking grill in case we wanted to cook out during the holiday weekend. By the time I finished adding items I thought we needed immediately, the truck was almost full. In the course of selecting and loading furniture and furnishings to get us comfortably through the weekend, I could see my mind-set shifting. For four years I had thought of the house as a dream to be constructed; now — it seemed almost suddenly — the house was a dwelling to be lived in. What would it be like? I knew from having watched it rise slowly out of the ground and acquire its exterior and interior finishes that the form of the house corresponded to my dream. It was in truth the fullfillment of what I had had in mind for so long. Would I find as much pleasure living in it as I had found in building it?

Lindsey helped me unload the heavy items when we got back to the new house. Then she and Laura drove to a restaurant we like to get some take-out dinner. Although our kitchen was fully operational, none of us felt like cooking dinner. We were all too tired and still too charged up. I finished unloading the truck and then rehung the bedroom dooors.

After dinner, Laura, Lindsey, and I sat in the family room on the sofa Lindsey and I had moved, watching television on the set installed in the corner cherry cabinet. The windows were open but all was quiet outside. (Throughout the long Fourth of July weekend I didn't hear a single firecracker.) The sense of privacy we had sought seemed attained. It had been a very long day, but I was much more conscious of a feeling of contentment than of tiredness. Most of our furniture was still in the old house, but where I was sitting was home. Emotionally, the move was already complete for me even though I hadn't yet — but soon! — slept in the new house. Everywhere

I looked I saw something my father or Bobby or I had built. My sense of the spiritual world may not be as vivid as Deanna's or Laura's, but I was comfortable thinking of the house as inhabited by the spirits of the many talented craftsmen and tradesmen who had built it. It was Lenny's fireplace as much as mine, Roger's terrace, Tedd's trusses, Joe's tile . . . As long as I dwell in this house, their work will not go unrecognized.

Suddenly there was an unfamiliar rustle. Laura, Lindsey, and I looked around to see what caused it. Blue, the large macaw, had been sitting on a perch outside his huge white cage in the all-seasons room. He must have been seized with an urge to join us, because he launched himself into the air and sailed across the ell past the kitchen toward where we were sitting. We didn't see the takeoff. His flight was smooth, his landing a little rough, but that could be excused for lack of practice. His gesture of joining us was a surprise ending to a long-awaited day.

My dream house was home. Not quite finished, but very ready to be lived in. If I dreamed any other dreams at all on that quiet and happy night, I don't remember them now.

Christmas 1994

 FOR HALF A YEAR now we've been living in the new house. That's long enough to notice what patterns will emerge in daily life in the house, and probably long enough to notice any major shortcomings in design or construction. All is well.

Structurally, the house looks very much the way it looked when we moved in just before the Fourth of July. I finished staining the clapboards and exterior trim his fall, but that's all I've found time for. It was as though everyone was waiting for me to occupy the house, and then I was immediately engulfed in a wave of demands on my time. Negotiations for the development of a magazine associated with *This Old House* required a great deal of time throughout the summer and autumn. I cut back on my personal appearance schedule as usual in July and August, but there wasn't any time for a vacation.

So the staircases are still plywood, and much of the interior finish woodwork remains undone. My unfinished office is filling up with various works in progress in a way that makes me wonder how I'll ever be able to clear it out enough to lay the wood floor, build and install bookshelves, and generally make it look civilized. Laura has been just as busy as I, necessitating a delay in furniture shopping.

With respect to design, only one minor flaw and one larger feature strike me now as things I might have done differently. The electrical contractor

The ell at night.

and I discussed installing a light switch in the kitchen that would be an alternate control for the lights on the exterior of the garage entrance, but it didn't get authorized or installed. It's a nuisance to have to go down to the garage to turn the lights on for security or to have to go back to the garage if they have been left on. The situation is one of those things that will bother me, I suppose, until I have it rewired.

The larger problem is the tight fit of the stairway in the main house. I still don't like the tightness of the curve at the bottom. Will Laura and I find it a feature that we consciously have to pay attention to as we use the stairs when we are older? Unfortunately, this is not an easy feature to change. The last riser of the stair at the second-floor hallway is about six inches from the opening for the guest bedroom door to the left of the stair, so there is no room to extend the length of the stair at the second floor. The hallway at the first floor is five feet wide, so when I framed the stairway I extended the last tread into the hallway — still leaving plenty of room for a comfortable hallway — hop-

Beyond Lindsey's cozy nook in the family room, the knoll rises to protect us from north winds.

In my unfinished but already fully operational office, I contemplate the view.

ing this would relieve the problem. It helped, but it still could be better, and I've used all the possible extra space available at the first-floor hallway. Changing the configuration of the stairs more than I've already done would mean redesigning the second floor. Perhaps, I think to myself, we should have made the house two feet wider; then we wouldn't have a tight curve in the stairway.

The mechanical systems work as well as I had hoped they would. The air-conditioning system kept us very comfortable on the hottest days of July and August. We have been fine-tuning the radiant heating system as the weather has gotten colder in a still mild winter. (How I wish we could have traded this year's November and December for the harsh months of 1992, when we cleared snow and ice away almost daily to raise the timber frame, install the roof, and complete the outside granite terraces.) We were familiar with central air-conditioning, in our old house, but our family had never lived with radiant heat. It provides a wonderful level of comfort with very quiet operation. About half the people visiting the house for the first time notice that there are no baseboard units or radiators, and the other half don't. I am very pleased with the absence of visible heating units.

From a construction standpoint, I think there's only one feature I would have done differently. The skipsheathing underlayment we used on the roof didn't permit the installation of an ice shield — a feature that keeps a buildup of snow and ice at the eaves from penetrating the space between shingles and underlayment. If we were doing the roof today, I would construct it almost the same way we did it in 1992. But on the edges of the roof I would replace the skipsheathing with four feet of plywood, then attach the gutters, then install an ice shield on the plywood, then apply the new matrix material to allow air to circulate between the underlayment and the shingles, and then, finally, apply the shingles. If I ever have to replace the roof while I'm living in the house, I will do that — unless, which is not unlikely, some even better system has been developed in the meantime.

When I lobbied Laura for our building a house rather than buying an existing house, I said that I could build her three times the amount of house and with workmanship satisfactory to me for the same money asked for the house Laura loved and I vetoed. Looking back, I see there was a bit of exaggeration

Lady, a Hawkhead, willingly responds to Laura's encouragement to flaunt the splendid crown of feathers on her head, particularly when a photographer is present.

in my claim. But we did build a little larger house and for substantially less money. Although I evaluated subcontractors' bids against my own experience for reasonableness, I selected them more for the quality of their work than for their price; and I always chose quality materials. Consequently, the construction is uniformly pleasing to me. Some features, such as the heating and lighting systems, are far beyond prevailing practices even in high-quality residential construction.

I haven't checked every invoice, but I've made enough of a calculation to see that the house cost us about $125 per square foot. A custom-built house with the quality and specifications of ours let out to a general contractor could easily cost half again as much per square foot. I haven't factored my own time into the calculation. The way I see it is that I did the house on an extended schedule that enabled me to keep up all my other professional obligations, without sacrificing opportunities or income. I did the house in my spare time. Therefore cost is not an element about which I have even an ounce of regret.

Time is the killer. I think back over the course of the construction, and I see what a sacrifice it involved for Laura and Lindsey for more than two and a half years. When I was home, which was too little for everyone's good — including my own — I wasn't home, because I was always up at the new house site checking on construction and doing work there myself. I wish I could have taken time off from all my other responsibilities just to build the house in one extended burst of activity. In my heart I know that wish is unrealistic. But I still find myself feeling deep regret that I couldn't frame the house myself. I don't feel any unhappiness about the framing John Conrad, Bill Delaney, and Woody Woodward did; they did a fine job. It's just one of those things I like to do myself. I get an immense sense of accomplishment in making a house take shape above the rough foundation.

Building a new house was a recovery of the way I used to work with my father. I didn't put any pressure on him to help me, and I think he was uncertain at first about getting involved. But once he did a few things, he really got into it. I never have to worry about the quality of his workmanship. We spent a lot of weekends together. A few of the tasks we did together — installing the upper kitchen cabinets, for example — could conceivably be done by one man with lots of time and ingenuity; but they're really by definition two-man jobs, and I had the right man working side by side with me. In the final few months before we moved in, my father's help was critical. We simply wouldn't have been able to move in when we did if he hadn't done a great deal of work hanging doors and doing other interior finish work.

Through the fall, as the weather turned cooler, I thought about the consequences of one pattern that had developed in life around the house. The double French doors from the great room of the ell out to the granite terrace had become the most used doorway by both family and visitors. My intention had been that visitors would park in the parking area next to the front walk and the old foundation and enter through the front door of the main house. Tom Wirth fully understood my intention. He put two features into the landscape design to encourage visitors to do as I wished. He proposed planting a group of trees and shrubs off the southeast corner of the main house that would screen the long south side of the house from view as a car came in the curving driveway toward the house. Visitors would have a full view only of the facade of the main house. He also proposed to pinch in the driveway just past the visitors' parking area, narrowing it a short distance to make it look more private before it widened again and continued past the house to the garage. Both these features would act as signals, he believed, to encourage visitors to pull into the parking area and approach the house by the front door.

I think his proposals will work, but we haven't had time to implement either of them yet. In the absence of an effective signal to use the guest parking area, everyone drives on back to the garage area. From there, the double French doors look like the guest entry. In warm weather I don't mind. But in cold weather opening those doors allows a blast of cold air into the great room. Better that visitors should enter the corner door from the same terrace into the mud hall just off the great room.

Until the Christmas season approached, I couldn't think of any way to encourage people not to use the double doors. Laura found a perfectly shaped ten-foot-high tree and bought about a thousand lights to decorate it. "Let's put it right in front of the double doors," I said, "where it can be seen best both indoors and from outside." And we did. My strategy is to keep the tree up through the twelve days of Christmas and maybe longer. Long

The main house at dusk.

enough that everyone gets accustomed to using a different door. The Christmas tree is just one of many holiday ornaments in the house. Laura has made the great room as beautiful for Christmas as any room in Victorian London could have been. Her loving decoration of our new home suggests that she is finding some of the deep satisfaction in living in our house that I found in building it.

Index

Throughout the index, page references in italics refer to captions; references in boldface type indicate pages where items are defined.